Biometric Technology

Authentication, Biocryptography, and Cloud-Based Architecture

Biometric Technology

Authentication, Biocryptography, and Cloud-Based Architecture

Ravindra Das

CRC Press is an imprint of the
Taylor & Francis Group, an **informa** business

CRC Press
Taylor & Francis Group
6000 Broken Sound Parkway NW, Suite 300
Boca Raton, FL 33487-2742

Printed on acid-free paper
Version Date: 20140926

International Standard Book Number-13: 978-1-4665-9245-2 (Hardback)

Visit the Taylor & Francis Web site at
http://www.taylorandfrancis.com

and the CRC Press Web site at
http://www.crcpress.com

This book is lovingly dedicated to Anita, my wife,
who has saved my life more than once.

This book is also dedicated to the loving memory of
my parents, Dr. Gopal Das and Kunda Das.

CONTENTS

Preface xvii
Author xix

1 An Introduction to Biometrics 1

Our Unique Features: Physiological and Behavioral 1
How Our Unique Features Can Positively Identify Us: The World of Biometrics 5
A Formal Definition of Biometrics 6
What Is Recognition? 7
Physiological and Behavioral Biometrics 7
What the Future Holds 8
The Granular Components of Recognition 9
Defining the Biometric Template 12
The Mathematical Files of the Biometric Templates 12
Understanding Some of the Myths Behind Biometrics 13
Understanding the Differences of Verification and Enrollment Templates 14
Summary of Points Covered 15
The Biometric Process Illustrated 16
For the CIO: What Biometric System to Procure? 17
For the CIO: Important KPIs to Be Examined 18
Looking beyond the KPI 22
U.S. Federal Government Biometric Metrics and KPI 23
Biometric Data Interchange Formats 23
Common Biometric Exchange Format Framework 26
Biometric Technical Interface Standards 28
U.S. Federal Government Biometric Testing Standards 30
A Review of Biometric Sensors 31
Optical Scanners 32
Solid-State Sensors 33
Ultrasound Sensors 34
Temperature Differential Sensors 34
Multispectral Imaging Sensors 34
Touchless Fingerprint Sensors 35
CCD Cameras 35
3-D Sensors/Review of Biometric Sensors in Use 36

The Disadvantages of Sensors 39
Typical Biometric Market Segments 40
Logical Access Control 40
Physical Access Control 41
Time and Attendance 42
Law Enforcement 43
Surveillance 45
Review of Chapter 1 48

2 A Review of the Present and Future Biometric Technologies 49

The Biometric Technologies of Today: Physical and Behavioral 49
Differences between Physical Biometrics and Behavioral Biometrics 50
Which One to Use: Physical or Behavioral Biometrics? 51
Fingerprint Recognition 52
The Unique Features 54
The Process of Fingerprint Recognition 55
Fingerprint Recognition Quality Control Checks 56
Methods of Fingerprint Collection 57
The Matching Algorithm 58
Fingerprint Recognition: Advantages and Disadvantages 59
Market Applications of Fingerprint Recognition 62
Hand Geometry Recognition 65
Hand Geometry Recognition: Enrollment Process 66
Hand Geometry Recognition: Advantages and Disadvantages 67
Vein Pattern Recognition 69
Components of Vein Pattern Recognition 71
How Vein Pattern Recognition Works 71
Vein Pattern Recognition: Advantages and Disadvantages 72
Palm Print Recognition 75
How Palm Print Recognition Works 76
Palm Print Recognition: Advantages and Disadvantages 76
Facial Recognition 76
Facial Recognition: How It Works 77
Defining the Effectiveness of a Facial Recognition System 79
Techniques of Facial Recognition 80
Facial Recognition: Advantages and Disadvantages 82
Applications of Facial Recognition 84
The Eye: The Iris and the Retina 86
The Iris 87

The Physiological Structure of the Iris 87
Iris Recognition: How It Works 89
The Market Applications of Iris Recognition 90
Iris Recognition: Advantages and Disadvantages 90
The Retina 93
The Physiology of the Retina 95
The Process of Retinal Recognition 96
Retinal Recognition: Advantages and Disadvantages 98
Voice Recognition 100
Voice Recognition: How It Works 100
Factors Affecting Voice Recognition 101
Voice Recognition: Advantages and Disadvantages 102
The Market Applications of Voice Recognition 103
Signature Recognition 104
The Differences between a Signature and Signature Recognition 105
Signature Recognition: How It Works 105
Signature Recognition: Advantages and Disadvantages 107
Keystroke Recognition 109
Keystroke Recognition: How It Works 110
Keystroke Recognition: Advantages and Disadvantages 111
Biometric Technologies of the Future 113
DNA Recognition 114
DNA Recognition: How It Works 114
DNA Recognition: Advantages and Disadvantages 116
Gait Recognition 117
The Process behind Gait Recognition 117
Gait Recognition: Advantages and Disadvantages 118
Earlobe Recognition 120
Earlobe Recognition: How It Works 120
Earlobe Recognition: Advantages and Disadvantages 121
Review of Chapter 2 122

3 For the C-Level Executive: A Biometrics Project Management Guide 125

Biometric Technology System Architecture 125
Sensing and Data Acquisition 126
Multimodal Biometric Systems 127
Single Sign-On Solutions 128
Implementing a Multimodal Biometric System 129
Challenges with Multimodal Biometric Systems 131

Signal and Image Processing 133
Preprocessing of the Biometric Raw Image 133
Quality Control Checks 135
Image Enhancement 136
Feature Extraction 137
Postprocessing 138
Data Compression 138
Data Storage 139
Search and Retrieval Techniques 140
Database Search Algorithms 141
Backup and Recovery of the Database 143
Database Configurations 144
Template Matching 145
Threshold Decision Making 147
Administration Decision Making 150
Biometric Templates Adaptation 152
Establishment of the Security Threshold Values 152
Reporting and Control 152
System Mode Adjustment 153
Privileges to End Users 153
Data Transmission 153
Biometrics Project Management 154
System Concepts and Classification Schemes 155
Upgrading to a Newer Biometric System? 157
The Feasibility Study 158
Application Classifiers 159
System Design and Interoperability Factors and Considerations 161
COTS-Based Biometric Systems 161
Proprietary or Open-Ended System? 162
Infrastructure Assessment 163
The Human Equation 164
Ergonomic Issues 165
The Feedback System 166
Population Dynamics 167
Systems Requirements Analysis 169
System Requirements Elicitation 170
System Requirement Analysis and Regulation 170
System Requirements Documentation 171
System Requirements Validation 172
Biometric System Specifications 172

System Architectural and Processing Designs 174
Storage and Matching Combinations 175
Operational Architecture Design—Multimodal Systems 177
The Information Processing Architecture 178
Subsystem Analysis and Design 179
Data Storage Subsystem Design Considerations 180
Database Management System 181
Determining the Security Threshold 183
Subsystem Implementation and Testing 184
System Deployment and Integration 185
Middleware 187
The Biometric Interface 187
End User/Administrator Training 188
System Maintenance 188
Upgrading an Existing System/Fine Tuning 189
System Reports and Logs 190
System Networking 191
Network Processing Loads 191
A Networking Scenario 192
Biometric Networking Topologies 192
Data Packets 193
Data Packet Subcomponents 194
Data Packet Switching 195
Network Protocols 195
TCP/IP 196
Client–Server Network Topology 197
Peer-to-Peer Network Topology 198
Routers 199
Routing Tables 200
Network Traffic Collisions 201
Review of Chapter 3 202

4 An Introduction to Biocryptography 205

Cryptography and Biocryptography 205
Introduction to Cryptography 205
Message Scrambling and Descrambling 206
Encryption and Decryption 207
Ciphertexts 207
Symmetric Key Systems and Asymmetric Key Systems 208

The Caesar Methodology 209
Types of Cryptographic Attacks 210
Polyalphabetic Encryption 210
Block Ciphers 211
Initialization Vectors 212
Cipher Block Chaining 212
Disadvantages of Symmetric Key Cryptography 213
The Key Distribution Center 215
Mathematical Algorithms with Symmetric Cryptography 216
The Hashing Function 217
Asymmetric Key Cryptography 217
Keys and Public Private Keys 218
The Differences between Asymmetric and Symmetric Cryptography 219
The Disadvantages of Asymmetric Cryptography 220
The Mathematical Algorithms of Asymmetric Cryptography 221
The Public Key Infrastructure 223
The Digital Certificates 224
How PKI Works 224
PKI Policies and Rules 225
The Lightweight Directory Access Protocol 226
The Public Cryptography Standards 227
Parameters of Public Keys and Private Keys 228
How Many Servers? 229
Security Policies 230
Securing the Public Keys and the Private Keys 230
Message Digests and Hashes 230
Security Vulnerabilities of Hashes 231
Virtual Private Networks 232
IP Tunneling 232
Mobile VPNs 234
Is the VPN Worth It? 234
Conducting a VPN Cost–Benefit Analysis 235
Implementing a VPN 236
The Components of a VPN Security Policy 238
End Users and Employees 239
The Network Requirements 240
Building Your Own VPN 241
Impacts to the Web Server 242
Impacts to the Application Server 243
Impacts to the Database Server 244

Impacts to the Firewall 245
VPN Testing 246
Implementing a VPN 249
Managing Public and Private Key Exchanges 251
The Access Control List 252
Internet Drafts 253
Four Vulnerabilities of a Biometrics VPN 255
Biocryptography 255
The Cipher Biometric Template 256
Biocryptography Keys 256
A Review of How Biocryptography Can Be Used to Further Protect Fingerprint and Iris Templates 257
Biocryptography in a Single Biometric System 257
Biocryptography in a Client–Server Biometric System 259
Biocryptography in a Hosted Biometrics Environment 261
Biocryptography and VPNs 261
IPSec 262
Review of Chapter 4 264

5 An Introduction to Biometrics in the Cloud 267

Introduction to Cloud Computing 267
The Basic Concepts and Terminology Surrounding Cloud Computing 268
The Cloud 269
Two Distinctions 270
The IT Resource 270
On Premise 271
Scalability 271
Asset Scaling 272
Proportional Costs 274
Scalability 274
Availability and Reliability 275
SLA Agreements 276
The Challenges and Risks of Cloud Computing 276
Security Risk and Challenge 277
Reduced Operational Governance 278
Limited Portability 278
Compliance and Legal Issues 279
The Functions and Characteristics of Cloud Computing 280
On-Demand Usage 281

Ubiquitous Access 282
Resource Pooling 282
Elasticity 283
Measured Usage 283
Resiliency 284
Cloud-Computing Delivery Models 284
Infrastructure as a Service 284
Platform as a Service 286
Software as a Service 288
Cloud-Computing Deployment Models 289
Public Cloud 289
Community Cloud 291
Private Cloud 291
Hybrid Cloud 292
The Security Threats Posed to Cloud Computing 292
Confidentiality 294
Integrity 294
Authenticity 294
Availability 294
Threat 295
Vulnerability 295
Security Risk 295
Security Controls 295
Security Mechanisms 296
Security Policies 296
Anonymous Attacker 296
Malicious Service Agent 296
Trusted Attacker 297
Malicious Insider 297
Traffic Eavesdropping 297
Malicious Intermediary 298
Denial of Service 298
Insufficient Authorization 298
Virtualization Attack 299
Overlapping Trust Boundaries 299
Important Mechanisms of Cloud Computing 299
Load Balancers 300
Pay-per-Use Monitor 301
Audit Monitor 301
Failover System 301

Hypervisor 302
Resource Clustering 302
Server Cluster 303
Database Cluster 303
Large Data-Set Cluster 303
Cloud Computing Cost Metrics and Service Quality Mechanisms 303
Network Usage 304
Server Usage 305
Cloud Storage Device Usage 305
Other Cost Management Considerations 306
Service Availability Metrics 307
Service Reliability Metrics 307
Service Performance Metrics 308
Service Scalability Metrics 309
Service Resiliency Metrics 309
An Introduction to Biometrics in the Cloud 310
Virtual Servers 315
Storage 315
Networks 316
A Detailed Example of How Biometrics in the Cloud Would Look Like 321
A Review of the Advantages and the Disadvantages of Biometrics in the Cloud 325

Index 329

PREFACE

Biometric technology has been around for a long time. For the most part, it has been accepted worldwide, without hesitation. It is a technology full of intrigue, mystique, awe, and misconceptions. Compared with other security technologies, biometrics receives the most scrutiny.

Why is this so? It is this way because it is a piece of us, whether biological or physical, which is being examined in close detail. Most populations are accepting of this fact, but some are not, especially in the United States. In our society, we view biometrics as a violation of our privacy rights and civil liberties.

However, truth be told, there are no magical powers behind biometrics. There are no *black-box* powers behind it either. It is just like any other security technology, with its fair share of benefits as well as flaws. Just like a computer program, it is literally *garbage in–garbage out*.

It is the goal of this book to dispel the myths and disbeliefs about biometrics as well as to review the strong benefits it has to offer to society, populations, and corporations. In addition, biometrics has a great future ahead of it in terms of research and development, which this book also reviews, by way of cryptography and cloud applications.

Although this book can be easily read and understood by just about anybody, it is primarily geared toward the corporate-level executive who is considering implementing biometric technology at their business or organization.

The author wishes to thank Mark Listewnik, John Grandour, Dennis Johnting, and Morgan Deters for the creation and development of this book.

AUTHOR

Ravindra Das was born and raised in West Lafayette, Indiana. He received his undergraduate degree from Purdue University, MS from Southern Illinois University, Carbondale, Illinois, and MBA from Bowling Green State University in international trade and management information systems, respectively.

He has been involved in biometrics for more than 15 years and currently owns a biometrics consultancy firm (Apollo Biometrics) based in Chicago, Illinois. He has been published extensively in Europe.

1

An Introduction to Biometrics

OUR UNIQUE FEATURES: PHYSIOLOGICAL AND BEHAVIORAL

From the very second we are born, there are a number of certainties we face as infants from the moment we come out of our mother's womb. First, there are death and taxes. Second, our DNA code is written with what our parents have faced or could possibly face in their lifetime. This includes certain ailments and diseases, such as coronary artery disease. No matter what we do to face these and try to overcome these hurdles, we will still be plagued with these issues as we grow older.

Another fact we face from the moment we are born is that we are all unique in each and every sense of the word from the rest of the 6 billion people who inhabit this planet. We are different from the next person in terms of our upbringing, our socioeconomic status, our personalities, our education, our interests, our goals, our chosen profession, the person we marry, you name it. Fortunately, for a fair amount of the world's population, we can enjoy certain amounts of comfort in life, especially in the Westernized and other developed parts of the world. However, there are many others to whom a life of despair and poverty is just a part of their everyday lives and unfortunately will be handed down to their next generations, especially developing areas in Africa and Asia.

However, wherever we originate from, there is another set of attributes that make us different from the rest of the world's population, and that is our personal physical attributes. What causes us to be so different from one another? Again, it goes back to that DNA code, which was written and coded into us from the moment of conception.

This DNA code comes from both of our parents, which affords us to possess these unique physical attributes. For instance, very often, it is quite remarkable to see how a child can so much closely resemble one parent, but look totally different from the other or when a child looks totally different from both parents. Realistically, no matter how close or distant we look from our parents, we still inherit their unique physical attributes. However, these are physical features that are visible and discernible to the naked eye.

As human beings, we also possess unique physical features that are different from everybody else on a much more microscopic level. What am I talking about? These unique features include everything from head to toe, literally speaking. From the top, our eye structures are different from everybody else.

For example, the retinal structure at the back of the eye (essentially, this is the group of blood vessels leading away from the optic disc going into the brain, which allows us to have vision) and the colorful iris at the front of the eye are deemed to be among the most unique physiological structures we possess. In fact, scientific research has even proved that these unique eye structures are different even among identical twins.

Obviously, even our faces are different from each other. We are not just talking about the visible physical appearance but also the geometry of the face. This includes the relative positions of the eye, nose, chin, ears, and eyebrows with respect to a fixed point on the face. Our voice structure has also been deemed to be different from others, as this unique trait is determined by the laryngeal pharynx.

As we move toward the center of the body, the most unique features are from within the hand itself. There are three main areas from which the hand possesses distinctive traits: (1) the geometry of the hand itself, (2) the fingerprint, and (3) the veins that exist from underneath the fingertip and the vein. With respect to the first unique trait, each individual's hand shape is different. This includes the relative positions of the fingers from the center of the hand, the knuckles, as well as the thickness, density, and the length of each finger.

The fingerprint has long been the focal point of major curiosity in terms of its own distinctiveness. On the outside of the tip of the finger, it hardly looks different, but when examined very closely at the microlevel, the differences become quite noticeable. The unique features are not in the ridges, whorls, or valleys, but rather they originate from the breaks and discontinuities that permeate from them.

These disjointed features are known as the *minutiae* of the fingerprint. There is no doubt that, for the longest time, the fingerprint has been associated with crime and law enforcement. A detailed history of how the fingerprint became such a powerful tool and association mechanism will be examined in this chapter.

Finally, just like the retina, there is a unique pattern of blood vessels from just underneath both the tip of the fingerprint and the palm itself. These are the veins, and although they are visible outside to the naked eye, the unique differences can only be seen at a microscopic level and under special lighting conditions. Although the discovery of these unique vein patterns is only recent, it is the minutiae of fingerprint that still predominates.

As human beings, we possess various behavioral characteristics that are different from everybody else's. If one were to look into the psychological books, one can probably find hundreds of different types of behavioral classifications. However, there are two primary ones that have captured attention and have proven their distinctiveness as well.

These two characteristics are the way we type upon a computer keyboard, and even more interestingly, the way we sign our name on legal documents. With the former, it has been determined that, for the most part, each individual has a different rhythmic pattern when typing on the keyboard. It does not matter if the text is long or short; there is a distinct uniqueness, no matter what the circumstances are.

With the latter, each individual also has his/her own distinct signature. However, keep in mind that it is not the signature itself that is the subject of interest, but rather, it is the way we sign our name that has garnered much attention. Internally, within our own bodies, we also possess unique features that separate us from everybody else. True, the exact mechanics and physiology inside our own bodies is a very complex thing to which, even today, scientists are still trying to fully understand.

However, it has been determined that physiological activities such as heartbeat, pulse rate, level of perspiration and sweat, and reaction to certain stimuli are unique for everyone. For example, while there is a textbook definition for what a normal heartbeat actually is, it varies greatly from person to person.

For instance, one person may have a very fast heartbeat, another may have a much slower one, and yet another may even possess a very irregular heartbeat, to which no baseline profile can be compared. These unique features from within and outside our own human bodies are the dominant ones known today, and scientific research has actually proven their distinctiveness.

However, as mentioned previously, as scientists are discovering more intricacies about the human body, more unique features are being discovered, at least tentatively (much more scientific research is needed to substantiate their distinctiveness). For example, our unique stride, or the way we walk in our daily lives, can be different from others (for instance, some individuals may walk in a zigzag fashion, a perfectly linear path, feet pointed inward or outward, rotate their hips differently, etc.); the shape of our ear (in a manner similar to hand geometry recognition) is unique from others as well as our body odor and the patterns of the sulci and gyri on the brain.

Given all these unique features we possess, it is not surprising that the idea of trying to use them to confirm the identity of an individual became of great interest. In the past, well into current times, there are multitudes of ways to see if an individual is who he or she really claims to be—for instance, passwords, personal identification numbers (PINs), challenge/response questions and answers, birth date or social security number confirmation, or just having sheer trust that the person is the actual he or she purports to be.

Although these methods have stood the test of time, there is one inherent flaw in all of them. That is, all of the above-mentioned techniques can be very easily circumvented. For instance, an individual can capture the password of another person (this is especially true with what is known as the *Post-it syndrome*, whereby employees write their password on a Post-it note and then attach it to their monitor's screen). Also, one can very easily guess the birth date, and through identity theft, even steal multiple Social Security numbers. Thus, if a person (namely the impostor who stole these numbers) were to call a financial center (such as a brokerage firm) and the operator asks for any bits of this information, this unsuspecting operator will be fooled into believing that the identity of the person on the other end has been confirmed.

The result is that the real person (the innocent victim in this case) will be drained off of his or her entire life savings. In this regard, at least, the challenge/response questions and answers pose a higher level of security. For example, unless you know the intimate details of an individual, the statistical probability of not knowing the right answer to the question posed increases much higher.

HOW OUR UNIQUE FEATURES CAN POSITIVELY IDENTIFY US: THE WORLD OF BIOMETRICS

Thus, given these inherent weaknesses, the next level *up* in terms of positively confirming the identity of an individual now comes down to these unique features just described. After all, if these features we possess are truly unique, would they not be proof enough to confirm our identity? In an absolute sense, this is true.

For example, if my fingerprint is unique, then the probability of me being who I claim to be is reasonably high, or if my iris structure is truly different from every individual on the face of the planet, then the statistical odds of being who I claim to be are increased tenfold (the reason for that is because the iris contains many more unique features than the fingerprint, and this fact will be explored in detail later in this book).

The science of positively identifying an individual based upon these unique traits is known as *biometrics*. However, keep in mind that this term is a general one. Meaning, it is also used in the medical field to describe one's physiological patterns (such as the heartbeat, pulse rate, the specific level of exercise that can be accomplished, etc.) and is also used in the market research world to gauge how a potential customer will react to certain advertisements, whether it is on the traditional television set or online.

However, based upon the context of this book, we are talking about biometrics from the perspective of the world of computer science. Meaning, unless one has the eyes that can see at a microscopic level, specialized technology is needed to positively confirm the identity of an individual. However, what kind of specialized technology are we talking about here?

At the most simplistic level, a capture device of some sort (such as a camera or a sensor) is needed to garner the raw image from which the unique features will be extracted and examined. Then, once this image is captured, certain mathematical algorithms are needed to filter out the unique features from the raw image, which was captured previously.

From this point onward, specialized software is then needed to transform these extracted, unique features into a certain *form*, which can be understood and used repeatedly by the technology. Then, subsequently, some sort of artificial intelligence at a primitive level is needed to compare the form that is stored by the technology for long-term (the *long-term form*) uses versus the forms that are stored for extremely short-term time frames (the *short-term form*).

Finally, at the end, the science of statistics is called upon to determine how close the long-term form closely resembles the short-term form, and if close enough, the person is then positively identified. All of this technology will be examined in much more detail later in this book. Although all these may look complex at first glance, this technology has greatly evolved over time, and the entire process of confirming the identity of a particular individual can take, literally, just a matter of seconds to accomplish.

A FORMAL DEFINITION OF BIOMETRICS

At this point, you may be wondering just what the exact definition of biometrics really is. If one were to conduct an exhaustive online search for its meaning, many different definitions appear, and this can be rather confusing. However, for purposes of this book, biometrics is specifically defined as "the use of computer science technology to extract the unique features of an individual, whether it be physical or behavioral traits, in order to positively verify and/or identify the identity of an individual, so that they may have access to certain resources."

When reading this definition, one of the questions that probably stands out is, *access to what resources*? Although the goal of biometrics is to positively confirm the identity of an individual, the result is to make sure that the right person gets access to the resources they are seeking to gain.

These resources can be anything from files that reside on a server from within a corporate intranet, to confidential patient files stored on a hospital server, to even gaining access to a secure room in an office building or even gaining access to the main entrance of a particular building. There are many other resources that a person can gain access to as well via the use of biometrics, but the ones listed above are the primary ones used extensively today.

Thus, to further illustrate our definition of biometrics, imagine that the computer science technology as just described is connected either to a main door or a primary (central) server. Once the identity of an individual has been confirmed via the technology, that particular individual will then gain instantaneous access to the files they need, or the door will open immediately, so that they can gain access to wherever they need to go.

Obviously, the reverse of all this is true as well. For example, a biometric system (as it will be referred to from this point onward) may even deny that that particular individual is **not** who he or she claims to be, and thus will be denied access to whatever resources he or she is requesting.

WHAT IS RECOGNITION?

In the world of biometrics, there is one term that is used very often, in fact, probably more so than other *lingo* in the vocabulary of biometrics, and that word is known as *recognition*. This means that it is the biometric system that is trying to recognize the individual in question, and confirm if he or she is indeed in the database. In fact, all of the biometric technologies are referred to as some sort of *recognition*.

The following is a list of all of the biometric technologies that are being used today and will be reviewed in detail in Chapter 2:

1. Fingerprint recognition
2. Vein pattern recognition
3. Hand geometry recognition
4. Iris recognition
5. Retinal recognition
6. Facial recognition
7. Keystroke recognition
8. Signature recognition

At this point, it should be noted that the first six biometric technologies are referred to as *physical biometrics* and the last two are referred to as *behavioral biometrics*. Why is there a differentiation between these two? If you think about it, the first six biometric technologies involve a physiological or biological feature of ourselves.

PHYSIOLOGICAL AND BEHAVIORAL BIOMETRICS

Physiological and behavioral biometrics means a snapshot that is being taken of the physical parts of our body, and then this particular snapshot (usually, more than one snapshot or picture is usually taken, and these are then aggregated together to form one overall, composite image) is then used to extract the unique features from which our identity will be confirmed. Hence, the term *physical biometrics* was conjured.

As mentioned previously, the last two biometric technologies form the group known as *behavioral biometrics*. It is still a part of our physical selves that is conducting the motion of typing on the keyboard and signing our name—in particular, our finger. However, if you were to look at a deeper level, there is no physiological component that is being caught.

Rather, it is the resultant motion of our fingers that is being captured. Let us look at this in a little bit more detail. For example, when typing on a keypad, there is a unique rhythm to it, especially in terms of how long we hold down the keys, the pattern or stride of our typing, the pressure we put upon the keys, how many times we use the same key, the speed at which we type our text, etc. However, anybody with a physical ailment (such as arthritis of the fingers) will have a much different typing rhythm and pattern when compared with an individual who does not have arthritis.

Moreover, when we sign our name, it is virtually the same type of behavioral characteristics that are being captured from our fingers. This also includes the pressure we put upon the pen, the angle of the pen when we sign our name, the speed at which we compose our signature, the type of fluid momentum involved (for example, do we pick up the pen at various intervals, or do we sign our name in one continuous motion), etc. It is very important to note that the actual uniqueness of the signature does not come into consideration at all here.

WHAT THE FUTURE HOLDS

As will be explored in Chapter 2, there is yet another group, or category, of biometric technologies, and this can be referred to as the *biometric technologies of the future*. There are three of these in particular that are worth mentioning:

1. DNA recognition
2. Earlobe recognition
3. Gait recognition

First, an important disclaimer needs to be made here. The biometrics in this list are not yet viable or proven technologies. Rather, they are all still undergoing a lot of research and development at this stage, and it will be a very long time until they establish their foothold in the marketplace. However, let us take a quick survey of these biometrics.

Concerning DNA recognition, if this proves to be a viable technology, it could even far surpass the uniqueness of the retina, which is now deemed to be the most unique biometric in terms of the richness of the data it possesses. Remember, our DNA code was written well before we were born and was coded into us when we were just a fetus and still evolving.

It is this DNA code that carries the *plan* for the rest of our lives. Thus, this fact gives credence as to the uniqueness and richness of the data the DNA strand (or code) possesses. However, the main drawback of DNA recognition is that it still takes a long time to thoroughly analyze. The days and hours it takes to conduct an analysis have to come down to just a matter of seconds in the world of biometrics.

In terms of earlobe recognition, the basic premise of the research and development being done here is to determine a way on how to accurately map the geometry of the ear, in a pattern very similar to that of hand geometry recognition. It has been determined that, at the very least, the ear possesses enough unique features, and the question now is how to map it from various locations (for example, mapping the distance between the ear canal and the earlobe).

Finally, with regard to gait recognition, as mentioned previously in this book, everybody has a unique stride, or in other words, a very distinct way of walking, that is different from everybody else. Some of the current technology being used to map our unique walking movements includes the use of radar.

If gait recognition does indeed emerge as a viable biometric technology, perhaps its greatest application will come in terms of airport settings, where the possibility of recognizing hundreds of individuals at once will indeed become a reality. Finally, concerning these potential technologies, it is gait recognition that holds the greatest promise, in terms of the progress of the research and development being made on it.

Finally, it is also important to note what biometrics is **not**. For example, although it was mentioned in this chapter that each individual possesses a unique heartbeat, pulse rate, perspiration rate, etc., which can be considered biometric traits as well since these are a unique feature, this view is from a medical standpoint only.

At present, there is no technology that exists (even at the research and development phase) that can accurately confirm the identity of an individual based upon heartbeat or the perspiration rate (even if such a technology were to exist, it would be deemed user invasive). Although these are unique physiological and biological events that occur in everyone, they cannot be considered a biometric from the perspective of confirming the identity of an individual. Remember, our definition of biometrics means finding those specific, unique traits that can be used to positively confirm who we claim to be.

THE GRANULAR COMPONENTS OF RECOGNITION

Now that we have turned our attention to our original definition of biometrics, there is one part that needs to be broken down into much further detail, and that component of the definition is the term *recognition*.

It is true that the biometric system is trying to determine who you are at a holistic level. However, there is a lot more that happens at the granular level of recognition. Thus, it is necessary to break down the individual components of recognition:

1. Verification
2. Identification
3. Authentication
4. Authorization

All biometric systems operate in two separate modes (or in both, if necessary): verification and identification, depending largely on the type of application that is being used and the scope of the biometric system that is to be deployed. It should be noted that most biometric systems operate in a *verification-only* mode.

These are the typical applications upon which you see on a day-to-day basis. These apps include single-sign solutions, computer/network access, physical access entry, and time and attendance. In reality, there are only a handful of applications that merit the use of identification.

The typical applications for identification applications include large-scale government projects, especially those focusing around law enforcement and the current war on terrorism. A perfect example of this is the AFIS database, which is currently being used by the FBI. AFIS stands for Automated Fingerprint Identification System. This is a gargantuan database that contains and holds the biometric fingerprint templates of all known or wanted criminal suspects.

Thus, for example, if a suspect is captured by law enforcement, that particular suspect can be linked to this database, to see if his or her fingerprint template exists in the AFIS database. This database is shared among three separate levels: (1) state level, (2) local level, and (3) international authorities, such as Interpol.

Now, turning the first component of recognition, which is verification, the question being asked is, *am I who I claim to be*? This means that we make a claim to the biometric system of our identity, and in turn, the biometric system has to confirm this claim by searching through all of the information and data that are stored in its database.

Thus, for example, when we approach a biometric system and present our fingerprint to the sensor or the camera, the unique features will be extracted and then compared to see if these features are very similar to the features that currently reside in the database of the biometric system.

Again, this is the typical operating mode of most biometric systems today in the corporate world.

With regard to identification, the primary question being asked is, *who am I*? This means that we are making no claims of our own, individual identity; rather, we are leaving it up to the biometric system entirely to determine who we actually are. In other words, it is not known if our unique information and data are already in existence in the database of the biometric system.

In these situations, the biometric system has to conduct an exhaustive search of its entire database to see if our unique information and data already exist from within it. Thus, again, for example, if we present our fingerprint to the sensor or camera, the unique features from it will be extracted and then used to see if we actually exist in the database of the biometric system.

However, a very important point needs to be made here. One may be wondering why verification is used much more often than identification. The answer to this is quite simple. Those applications that use verification can confirm the identity of an individual in typically less than 1 s, whereas those applications that use identification can take much longer, typically minutes or even hours to confirm the identity of a particular person. In addition, verification applications consume much less network and bandwidth resources than those applications that use identification.

Once the verification and/or identification stages have been completed by the biometric system, the next two steps of authentication and authorization occur very quickly, in rapid-fire succession. Authentication simply means that the unique features that have been extracted have been indeed confirmed and accepted by the biometric system (meaning those information and data already exist in the database).

Authorization simply means that after we have been accepted by the biometric system, we now have permission to access whatever resources we are requesting to use (such as access to confidential computer or network files or access to a secure room in an office building by having the door open electronically, etc.).

As discussed in this chapter, after the raw images of our physiological and biological features are captured, the unique features from these images are then extracted. For instance, from a fingerprint image, the minutiae (which are the breaks and the discontinuities in the ridges and the valleys) are what is captured; from an iris image, the spatial orientations of the unique features will also be captured; and from a facial image, the relative positions of the eyes, the eyebrows, chin, nose, etc. will be captured.

DEFINING THE BIOMETRIC TEMPLATE

These unique features are then transformed into a special type of *form*. There is a specific name for this *form*, and it will be used throughout the remainder of this book. This form is known as a *biometric template*. The specific definition of a *biometric template* is a digital representation of the unique features (either physical or behavioral) of the raw image that is captured from the individual in question.

However, biometric templates are a little bit more complex than that. Rather, they are mathematical files, and the exact type of mathematical file depends upon the specific biometric that is being collected, and of course, the biometric vendor who developed the exact biometric system being used to collect the raw images.

Let us illustrate this with an example, and again, we will turn to the most basic one of that of the fingerprint. After the minutiae are collected, examined, and extracted, they are then converted into a mathematical file known simply as a *binary* mathematical file. We all know that binary digits are simply a series of zeros and ones. Thus, as you can imagine, the fingerprint template now becomes this series of zeros and ones in the biometric system.

Thus far, after reading this chapter, you will notice a sequence of events occurring, which can be termed as the *biometric process*. In this process, we have examined the presentation of the biological and physiological features being captured by the sensor or the camera, followed by multiple images of this feature being captured and being converted into one composite image, and from there, the unique traits of that feature being captured and analyzed and converted into the biometric template.

THE MATHEMATICAL FILES OF THE BIOMETRIC TEMPLATES

It should be noted that it is this biometric template, or mathematical file, that is stored into the biometric system. The raw images of our physiological and biological features are never stored permanently in the biometric system. If they are stored, it is for a very short time only, up to the point only when the unique features are extracted, and then these raw images are subsequently discarded.

As mentioned earlier, each different biometric technology possesses its own style and kind of mathematical file. The following is a representative sampling of some of these files:

1. *Fingerprint recognition*: Binary mathematical files are used.
2. *Hand geometry recognition*: Binary mathematical files are used.
3. *Iris recognition*: Traditionally, Gabor wavelet theory mathematics was utilized, but with the expiration of the original patents of Dr. John Daugmann, many other types of mathematical files have emerged onto the marketplace.
4. *Facial recognition*: Eigenfaces and eigenvalues have also been the traditional mathematical files used, but with the advancement in technology, other types of mathematical files have come out as well (again, this will be explored in much more detail in Chapter 2).
5. *Keystroke recognition/signature recognition*: Since these are behavioral biometrics, there are no physiological or biological characteristics that are being captured. Rather, statistical modeling is very often used, and from there, statistical profiles are generated, which then become the respective biometric templates. It is the theory of hidden Markov models that is used the most frequently.

UNDERSTANDING SOME OF THE MYTHS BEHIND BIOMETRICS

Even among the public, there is a common myth that exists about biometric templates. In fact, it is this very myth that triggers the biggest fears about biometrics, and subsequently, its very slow adoption rate in the United States. This myth is the misconception that it is the raw image that is stored, when in fact, only mathematical data reside in the database of the biometric system.

As a result, the question that gets asked too many times is: What if my biometric template is stolen? Is that the same as credit card theft? This is true, but only to a very small degree. Any type of hacking, whether biometric template or not, is still theft, but it is nowhere near the level of what the damage credit card theft can do.

After all, since the biometric templates are just mathematical files (and the algorithms that support them), which are unique to each specific technology and vendor, in the end, what can a hacker do with them? Nothing, really, in the end. For example, it is now likely that a hacker can take these biometric templates and purchase high-end merchandise, as opposed to accomplishing this task with a stolen credit card number.

UNDERSTANDING THE DIFFERENCES OF VERIFICATION AND ENROLLMENT TEMPLATES

Although we have examined in detail what a biometric template exactly is, there are two specific types of templates that are used throughout in the entire biometric process. Both of these templates are known respectively as the *enrollment template* and the *verification template*. Theoretically, there is nothing different about these two types of templates; they are just merely mathematical files, just as described previously.

However, the only real difference between these two templates is at the time the unique features are recorded into them and the purposes they are used for. With the enrollment template, this is the first biometric template that is generated when an individual enrolls into the biometric system for the very first time, as a new, registered user. Also, this is the template that gets stored permanently into the database of the biometric system (unless of course the systems administrator decides for whatever reason to delete this particular template).

Of course, it is safe to assume that the same user will be using this biometric system on a regular basis to get access to whatever they need to get to (such as confidential computer files or access to a secure room). Every time this particular individual uses this very same biometric system, he/she will have to present his/her biological or physiological feature component again.

When this happens, the unique features still get extracted, but this time, these unique features get converted into what is known as the *verification template*. This particular template is then compared with the enrollment template, and then a decision is made as to how closely these two templates *resemble* each other. If they are close enough from the standpoint of statistical analysis, then the individual is positively verified.

However, if these two templates are not close enough in resemblance, then this particular individual will then be denied by the biometric system, and as a result, he or she will be denied access to whatever resources they are requesting at that particular point. It should also be noted that the verification template is typically discarded from the database at this point, since there is no need to store it permanently (this is based on the assumption that the individual will be using the same biometric system over and over again, until they are no longer affiliated with that particular place of business or organization).

Let us illustrate this entire process with an example, this time using iris recognition. Suppose that a business has just implemented a brand new iris recognition system; the employees whose irises have been scanned for the first time will become the enrollment templates and will get permanently stored into the database of the iris recognition system. As these same employees come back to their place of work everyday, their irises will be scanned on a daily basis.

These biometric templates that get recorded daily subsequently become the verification templates and will get compared with the enrollment templates that have been permanently stored. If there is enough statistical closeness between these two templates, the employees will be allowed to enter the place of business, and if not, their identity will then have to be confirmed by other, more traditional mechanisms.

A question that often gets asked at this point is if there is such a thing as two identical biometric templates (in other words, can there be an enrollment template and a verification template that are duplicate in nature from the same individual?). The answer is no. Biometric templates are never 100% the same; thus, that is why the statistical similarity of the enrollment template and the verification template is examined, i.e., to see how closely they match or not match.

There are many reasons why these two templates are never 100% the same, and much of it can be attributed to the environmental conditions, the external lighting, or the way a particular individual presents his/her physiological and biological traits to the sensor or camera.

SUMMARY OF POINTS COVERED

In summary, thus far, we have covered a lot of material. To review, we have defined what biometrics is all about and what it is not; examined what verification, identification, authentication, and authorization are; and covered the various types of biometric templates. With all of these concepts now covered, we can now put together the pieces and present a complete, yet very basic biometric process.

It is important to keep in mind that this process, the way it is presented, is very simple. Most, if not all, biometric system deployments and installations in the real world are much more complex and have many more components especially in the way of networking, the databases, and the associated software applications that go along with them. However, it is this very basic biometric process that forms the backbone of all these procurements and deployments.

THE BIOMETRIC PROCESS ILLUSTRATED

Thus, let us illustrate this biometric process with facial recognition. Again, imagine that a business has just deployed a facial recognition system. The employees have started to register themselves into this, and to capture the best, most unique features possible from the face, multiple raw images are captured and then compiled into one, primary image. This is known (even throughout all of the biometric technologies) as the *composite image.*

It is from this composite image that the unique features are extracted, and from there, converted into the enrollment template, which is the template that gets stored permanently into the database. This part of the biometric process is known as *enrollment* because individuals are literally enrolling into the biometric system for the very first time.

Then, when the same employees come back to the place of business the next day, to gain access to their workplace, they have to register again into the facial recognition system. Multiple raw images are again captured; the unique features are extracted and then converted into the enrollment template. As reviewed, the verification template is then compared with the enrollment template to ascertain the statistical closeness of these two templates.

In the world of biometrics, it is the matching algorithm that actually does all of these calculations and sends back the result to the biometric system, and then in turn, to the employee trying to gain access. Keep in mind that the matching algorithms are proprietary from one biometric vendor to another and are literally built into the biometric system.

There is nothing that can be done to modify or alter these specific matching algorithms. The only thing that can be done by the systems administrator is that the specific threshold of the *sensitivity* (in other words, how close the templates really need to match in order to accept legitimate users and deny illegitimate users entry or access) of the matching algorithm can be modified or adjusted to best fit the security needs of the place of business. The *sensitivity* of the matching algorithms is determined by a set of biometric key performance indicators (KPIs).

Once the verification template and the enrollment template are compared and the degree of closeness is determined, the biometric system will then either accept or reject the employee. Of course, if these two templates are deemed to be statistically close enough in nature, the employee is then accepted by the biometric system. This entire process of comparing these two templates is known as *verification.*

Of course, illegitimate users will be denied by the biometric system. In addition, remember that biometric systems are not infallible; they are prone to errors as well, and even legitimate users can be denied as well. Finally, as reviewed, once the employee has been through the verification by the facial recognition system, the steps of the biometric process are authentication and authorization, and the employee will then be given access to whatever resources that they have requested.

Although the biometric process sounds like a lot of steps to be accomplished, it is all accomplished in just 1 s or less. This biometric process is illustrated in the following figure.

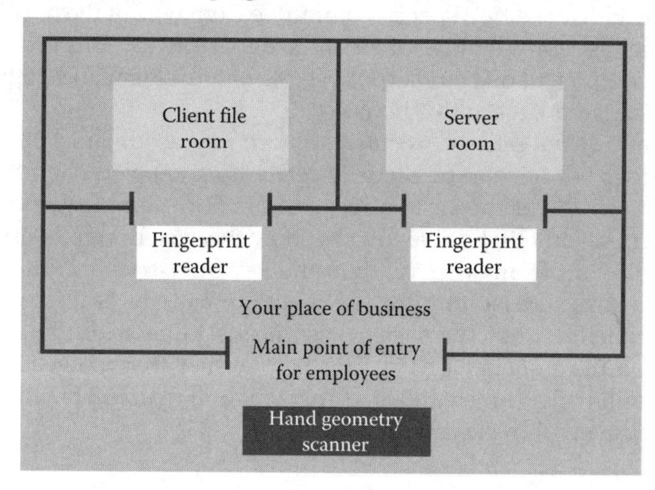

For the CIO: What Biometric System to Procure?

One of the most important decisions a business owner or a C-level executive can make is in deciding what type of biometric system to acquire for their place of business. There are a whole host of factors that need serious consideration in this decision-making process. Some of them include the exact security needs of the business (for instance, will biometrics be used as the sole means of security or will it be added onto an existing security infrastructure?); the results of the systems analysis and design that should be conducted before any procurement decisions are made; the applications of which the biometric system will serve (for example, single sign on, physical access entry, time and attendance, network access, etc.); what kind of technology and support the chosen biometric vendor can provide; etc.

However, obviously, it is the goal of the C-level executive to procure a security system that will prove itself in terms of a positive rate of return on investment (ROI). Often, trying to calculate the ROI of a biometric system deployment can prove to be a very complex task, again depending on the size and the magnitude of the project in mind.

However, calculating the ROI for a very simple biometrics should prove to be very easy to calculate. At this level, we are simply looking at a stand-alone biometric system (such as one device) at a small business location (perhaps 20 employees or fewer). However, keep in mind that all of these variables just posed are actually project management issues that need to be addressed by a special committee from within the organization itself, which is representative of all of the departments. This project management component of a biometric systems procurement and deployment will be discussed later in this book.

However, as a C-level executive, before any acquisition of any new system (does not have to be security related per se), you need to look at some sort of KPIs of the system in question. For example, by looking at these metrics, you will be able to know how the system stands up to comparable solutions from other vendors and key industry benchmarks.

It is always so easy to purchase a system as long as it is within the budget guidelines, but the system also needs to be critically examined based upon the associated KPIs. This is to ensure that the needs of your business will truly be met, and also, that you will realize a positive ROI in the end, with all of the investments made.

For the CIO: Important KPIs to Be Examined

In the world of biometrics, it is so easy to acquire a biometric system just because it looks *good*, or is within budget, without even looking at some of the top KPIs that relate to it. Thus, at this point, it is very important to examine and review some of these KPIs, as they relate to a procurement decision of a potential biometric system. Some of these top biometric KPIs include

1. The false acceptance rate, also known as the FAR or type II errors
2. The false rejection rate, also known as the FRR or type I errors
3. The equal error rate, also known as the ERR
4. The ability to verify rate, also known as the AVR
5. The failure to enroll rate, also known as the FER

With regard to the FAR, this metric reflects the possibility of an illegitimate user, or even an impostor, being accepted by the biometric system. For example, this happens when John Doe, who is not registered into the biometric system or even does not have any affiliation with the place of business, for some reason or another, is actually verified by the biometric system, is allowed into the business, and is given access to confidential network files he has no business or reason to be into.

The above example actually does happen, and as mentioned previously, biometrics are not infallible and are prone to faults and errors just like any other technology.

With respect to the FRR, this metric reflects the statistical probability of an individual who is legitimately enrolled into the biometric system actually being denied by it. In other words, for example, suppose John Doe is a legitimate employee of a particular place of business and is enrolled into the biometric system. For some reason or another, despite his legitimacy, he is totally rejected by the biometric system and is denied access to entry or whatever resources they have requested.

In terms of the ERR, this is where the FAR and the FRR equal each other, and thus is the ideal, or optimal, setting for any type of biometric system that is to be deployed.

The AVR describes the overall percentage of a particular population that can actually be legitimately enrolled into the biometric system. For instance, it does not matter what type the exact population is. It can be the total number of employees in a place of business or the entire citizenship of a particular country. All that matters is the total number of people that can be properly enrolled into the biometric system.

In terms of mathematics, the AVR can be thought of as the union, or the combination of the FER and the FRR. Specifically, the formula is

$$AVR = [(1 - FER) * (1 - FRR)].$$

Finally, the FER statistically describes that percentage of the population that **cannot** be legitimately accepted to the biometric system. This metric can also be thought of as the converse, or the mirror image, of the AVR. There are a number of reasons why particular individuals cannot be enrolled into a biometric system.

For example, the most likely causes are that of physical ailments, such as arthritis, skin discoloration, blindness, and the sheer lack of physical and biological features.

All of these metrics of the FRR, FAR, and ERR are depicted in the following figure.

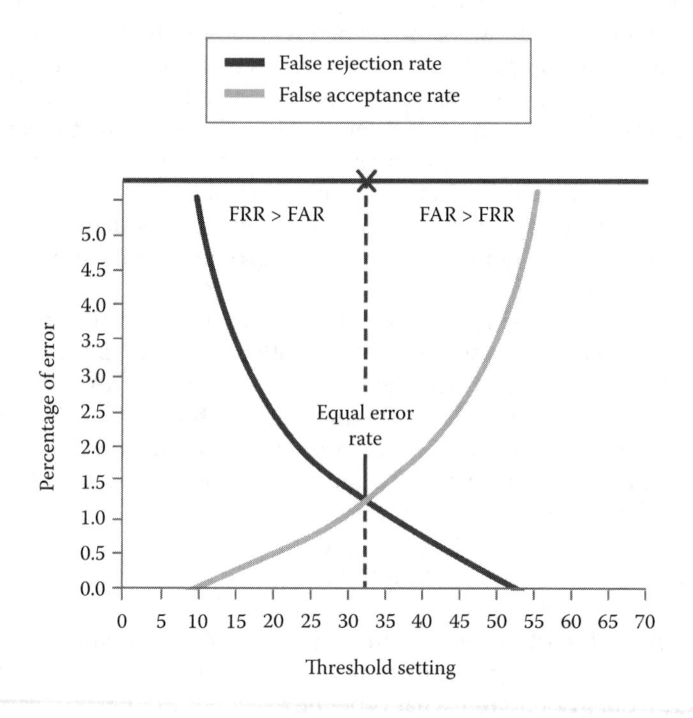

Upon examination of the aforementioned figure, the y-axis shows the percentage of error and the x-axis displays the threshold setting of the biometric system that the systems administrator can establish on the biometric system (which was alluded to when we examined enrollment templates and verification templates).

The black curve represents the FRR, and the gray line represents the FAR. In the middle is the ERR, which is, in theory, the optimal setting of a biometric system.

The region to the left of the ERR shows where the FRR is greater than the FAR, and the region to the right of the ERR shows where the FAR is greater than the FRR. With the former, this scenario displays where more legitimate users are being rejected than illegitimate users being accepted, and with the latter, this scenario shows where the number of illegitimate users being accepted is greater than the number of legitimate users being rejected by the biometric system.

Obviously, as one can tell, being at the very center is the best condition for a biometric system. However, again, this is only a theoretical dream to be achieved, as getting to this level is almost impossible in the real world of biometric deployments and installations. It should be noted here that even if a biometric system is the simplest in terms of systems analysis and design (meaning, just one stand-alone biometric device is being used at the place of business), trying to achieve the ERR is virtually impossible.

Thus, where is the next best position for a biometric system to be at according to this graph? For obvious reasons, you do not want a high level of illegitimate users being accepted into the place of business by the biometric system. Realistically speaking, the next best level to be at is the region where the FRR is greater than the FAR.

Thus, at this point described, the number of legitimate users being denied access by the biometric system is much greater than the number of illegitimate users being given access. At least in this scenario, you have a higher probability of illegitimate users being denied access and a somewhat lesser number of the legitimate users being granted access, which is a much better scenario.

Ultimately, you want to have zero illegitimate users being accepted and zero legitimate users being denied access by the biometric system, which is another way of describing the ERR.

Finally, an appropriate threshold setting based upon the scale of the graph would be somewhere between 25 and 30, which will result in a much lower percentage of errors being given by the biometric system.

As mentioned in this chapter, the threshold can be established by the systems administrator of the biometric system when comparing the enrollment templates against the verification templates. The threshold setting just described above is exactly what we were talking about. That threshold setting is where the biometric system needs to be at, so one can achieve the point where the least number of illegitimate users will be accepted and the least number of legitimate users will be denied access by the biometric system.

LOOKING BEYOND THE KPI

The biometric metrics just described above are those that have been set forth from within the biometrics industry itself. These are not just recently developed metrics. Rather, they have been developed over a long period and are not just used in the United States; they are also very much used on a global basis by other biometric vendors. For example, when one attends a conference or other such related venue regarding biometrics, these metrics are one of the most talked about topics amid all of the other technical and social aspects surrounding biometrics.

However, it is also important to keep in mind that these metrics just discussed have no federal government oversight to them. In other words, there is no enforcement body or regulatory agency in place to ensure that the claims made by biometric vendors with regard to these metrics when it comes to their products and solutions are actually misleading or not.

So, as a C-level executive, it is very important that as you go through the decision-making process of the evaluation, procurement, and acquisition of a biometric system for your place of business, you fully evaluate these metrics carefully.

Although most biometric vendors are honest about their claims, this is still a very competitive industry, and all of the vendors whom you are examining for your purchasing decision will obviously present their product or solution in the best light to you. As the C-level executive, if you are unsure about the specific questions to be asked, there are always vendor-neutral biometrics consultants whom you can refer to in order to get a second opinion as to the claims made by these biometric vendors with regard to these industry-wide metrics.

After all, the bottom line is that you are trying to protect or enhance the level of security at your place of business; therefore, it is very important that you make the wisest decision possible, even if it means turning to one of these consultants to guide you through understanding what these metrics mean to your specific security needs.

Although the federal government in the United States provides no oversight to these industry metrics, just within the last decade or so, they have become the biggest customer of biometrics, far surpassing even that of the private sector by tenfold in terms of biometric procurements and acquisitions.

As a result of this huge buying activity, the federal government has also become the largest awarder, in terms of dollar volume, of these lucrative biometrics contracts. To make sure that every dollar and cent is well spent, the federal government has even created its own set of standards and metrics by which to ascertain the effectiveness and legitimacy of the various biometric technologies that are available today as well as their interoperability with other systems and deployments.

U.S. FEDERAL GOVERNMENT BIOMETRIC METRICS AND KPI

Given the gargantuan size of our federal government, it is the many types of agencies that exist from within it that use all of these biometric technologies as well as the whole gamut of applications that are available. Because of this, there are literally hundreds of biometric standards and metrics that have been developed as well as the many committees and subcommittees that have been established to formulate these biometric standards and metrics.

It is totally out of the realm and range of this book to review each and every standard and metric set forth by the federal government, but it is very important for the C-level executive to gain insight into some of the most important ones that also have a strong bearing upon the commercial market and the private sector.

Thus, the standards and metrics that have been selected for further review include the following:

1. Biometric data interchange formats
2. Biometric data structure standards, which include the common biometric exchange format framework, also known as CBEFF
3. Biometric technical interface standards, which include the Biometrics Application Programming Interface, also known as BioAPI

Biometric Data Interchange Formats

With regard to the first grouping of standards and formats, these are standardized methodologies that permit for the interaction of various biometric templates that are using the same type of biometric technology. To illustrate, imagine a fingerprint biometric system. Suppose that this system is installed at a Fortune 500 organization, with global offices located in many countries. This organization does not have just one type of fingerprint recognition technology from just one biometric vendor.

Rather, they have multiple kinds of fingerprint technologies from various fingerprint recognition vendors. It is the biometric data interchange formats that allow for these various fingerprint readers to communicate with one another and that allow for the free flow of biometric information and data among them. Put another way, it is this interchange format methodology that allows for the free communication between various devices of the same type of biometric technology.

Keep in mind that these technologies do not have to come from the exact same vendor; they can be from various ones. However, the key point is that the biometric modality itself (another term for a specific piece of biometric technology) must be the same for this methodology to truly work. More specifically, the goals and objectives of the biometric data interchange formats include the following:

1. Having a commonality or standardization among biometric templates, which includes a common format for representing the biometric templates; the allowance for the smooth and free flow of biometric information and data (this can also be referred to as the *transfer syntax*); and having the ability for platform and biometric vendor independence
2. Providing clear guidance of what type of headers the biometric templates are using (this can also include the metadata, which is essentially data about the data)
3. Promoting new ideas with regard to the research and development of the biometric sensor types and their feature extraction abilities
4. Permit for the successful exchange of biometric information and data, which is measured by the following variables: seamlessness, correctness, and effectiveness

In addition to the above goals and objectives, the biometric data interchange formats have specifications for the following:

1. Having a standardized format for incoming and outgoing biometric information and data, which include the actual biometric templates and their associated raw images as well as any type of raw signals.
2. Having an open-source model in which each type of biometric technology from each specific biometrics vendor and each type of biometric template associated with that particular technology as well as their related, major biometric information and data subsets can communicate and be interoperable with each other.

It is the interoperability that is key here, and it should be noted as well that more biometric information and data subsets can greatly diminish the interoperability function between the various technologies (of course, again, the same modality is used) among the various vendors.

From within the level of the federal government, there are three primary working groups that are currently working on this project. This includes the following:

1. ISO/IEC JTC 1 SC 37 WG 3-19794: This group actually develops the biometric information and data interchange format standards. In fact, their standards have been adopted by the International Civil Aviation Organization (ICAO) in the development and implementation of the e-passport.
2. ANSI/NIST ITL for Law Enforcement: This group developed the standards and protocols as described above, primarily for law enforcement application. An example of this is for the exchange of high-resolution, grayscale fingerprint images among the various fingerprint databases, which include those of AFIS and the fingerprint databases possessed by Interpol.
3. ANSI INCITS Data Format Standards: This group has developed standards that encompass the gamut of biometric technologies including
    a. Fingerprint recognition
    b. Iris recognition
    c. Facial recognition
    d. Signature recognition
    e. Hand geometry recognition
    f. Voice recognition
4. Other projects have included
    a. Developing standards for biometric sample quality (this deals with the multiple images, which are taken to form the main composite image, as discussed in this chapter).
    b. Fusion data: This standard has been created in an effort to harmonize the comparison score results when the enrollment templates are compared with the verification templates.

When computer systems and applications communicate with another via data packets, they use various data fields (which is the information about the content of the data and what it specifically holds) to fully understand how to process the certain types of packets that are being transmitted back and forth. However, if for some reason, these data fields do not match up from where they originate (such as the computer or application), important decisions about what to do next in the process or hierarchy may simply go unnoticed between the systems.

This situation can also be extrapolated into the world of biometrics. For example, if two hand geometry scanners are communicating with another and the data fields about the biometric templates for some reason do not match up properly, the systems may not work in the sense that the all important positive verification and identification of a particular individual may never even occur.

In these types of scenarios, the only option is to rewrite the software programming code in both of the hand geometry scanners. Obviously, this work can be very time consuming as well as an expensive process for a place of business to embark upon. Thus, for this, the use of logical data standards comes into play here, which is the second major grouping in the list presented earlier.

Common Biometric Exchange Format Framework

With the use of logical data structures, the biometric templates are literally placed into a *virtual wrapper*, which converts the data fields into a universal language that both hand geometry scanners can recognize very easily, based on the above example. Therefore, in biometrics, it is the Common Biometric Exchange Formats Framework, or CBEFF, that provides the foundation for this second major grouping of standards. Specifically, it can be defined as "a standard that provides the ability for a system to identify, and interface with, multiple biometric systems, and to exchange data between system components" (*Certified Biometrics Learning System, Module 4, Biometrics Standards,* © 2010 IEEE, pp. 4–52).

It should be noted here that the CBEFF standards do not provide for the actual ability of biometric systems to communicate with one another; rather it provides the means so that the biometric templates (and any other relevant data or information about them) can be more easily understood by the biometric devices that are being used.

While the first major grouping of standards, the biometric data, interchanges formats, CBEFF allows for the free flow of the biometric templates in a heterogeneous environment. Meaning, different biometric modalities made from different vendors can now easily communicate with one another (for instance, a hand geometry scanner communicating with an iris recognition scanner). CBEFF consists of the following properties, which permit for this free flow of biometric templates:

1. *Data elements*: A list of the metadata elements has been developed so that the biometric templates (and their associated information and data) can be exchanged with multivendor hardware, software, systems, and subsystems.
2. *Common file format*: With these file formats, a certain biometric technology can work with another biometric technology not just at the template level but also at the hardware and software component levels.
3. *Extensibility*: This sophisticated feature of CBEFF allows for the quick and easy development of new data file formats, registers these new file types, and provides identification markers for them from within the biometric software code itself.
4. *Data formats*: Anybody (such as a software developer or biometrics vendor) that registers new data file types as described above will be issued a timestamp, version history number, expiry data for the newly registered data files, and the name of the creator of the file types. From this information, other people and organizations can then easily access the data and use them for their own biometric systems, if the need arises.
5. *Biometric information records (BIRs)*: The data file types that have been created have to conform to a CBEFF format, so that the biometric templates and other relevant information and data can be easily exchanged between systems.

Despite the advantages that the CBEFF standards have to offer, it has one serious disadvantage. That is, while these particular standards allow for the free exchange of biometric templates and their much easier interpretation between heterogeneous systems to occur, CBEFF in no way guarantees automatic interoperability.

The bottom line is that despite these standards, the various biometric templates are still truly compatible only with those biometric devices and applications that support them. Finally, CBEFF is composed of two other standards:

1. *ISO/IEC 19785*: This applies specifically to nonbiometric, information technology applications.
2. *ANSI/INCITS 398-2008*: This spells out the goals and objectives for CBEFF.

Biometric Technical Interface Standards

The third major group of standards, known as the BioAPI, specifies the level of interchangeability of biometrics hardware, software, middleware, systems, and subsystems. Specifically, it can be defined as follows: "[It] ... defines the application programming interface and service provider interface for a standard biometric technology interface. The BioAPI enables biometric devices to be easily installed, integrated or swapped within the overall system architecture" (*Certified Biometrics Learning System, Module 4, Biometrics Standards*, © 2010 IEEE, pp. 4–68).

A number of key features of the BioAPI need to be noted at this point. First, the BioAPI standards deal with the integration of the various biometric modalities. It has nothing to do with the interaction of the differing biometric modalities. By integration, we mean how easily can a new biometric device be installed and configured into an existing biometric system. For example, if a hand geometry scanner were to fail, how quickly can a differing biometric modality, such as iris recognition, be installed and assume the responsibilities the hand geometry scanner once held?

Thus, one important function of the BioAPI standard is what is known as *plug and play*. Essentially, this means how quickly a new biometric technology can be installed (which is the plug portion) and configured to work in an existing environment (which is the play portion).

Second, the BioAPI standards are designed to work in an open source, software model. Meaning, any biometric software code that is developed under BioAPI can be easily shared with other software developers and biometric vendors. One of the primary goals of the BioAPI standards is to create an environment where people who are involved in developing biometric applications have the ability to work independently in a vendor-neutral environment.

The BioAPI standards provide a number of key benefits, which include the following:

1. It fosters an environment where all of the components of a biometric deployment can be easily interchanged with other, differing components.
2. Any biometric solution created by a biometric vendor is made to be *generic* enough so that it can support the *plug and play* benefit and work/operate with virtually any other biometric modality.
3. Any other security system that is nonbiometric in nature can be easily added onto biometric-based solution.

4. Biometric-based applications can be developed very quickly and easily.
5. Biometric solutions can become platform- and operating system–independent of one another.
6. Differing and multiple biometric modalities can be literally *fused* together into one harmonious environment.

The BioAPI standards are composed of the following components:

1. *ANSI INCITS BioAPI Standards*: This specifies the subcomponents of the BioAPI; to support biometric modality fusion (described in Item 6 above); the requirements for ten fingerprint capture and unique feature extraction; and the requirements needed to ensure that a nonbiometric solution can be easily integrated into a biometric-based solution.
2. *ISO/IEC 19784*: This specifies how biometrics-based software solutions graphical user interface (GUIs) should be designed; the use and exchange of security certificates within biometric systems; the means to archive biometric information and data; and the operation of a multiuse interface for biometric sensor communications and interoperability.

Currently, it is BioAPI 2.0 (this supersedes BioAPI 1.1) that is being used across the board for the development of biometric-based applications and solutions. Some of the new features of the BioAPI 2.0 include

1. Software component and subcomponent calls
2. A definition set of high-level software functions
3. A definition set of primitive (or lower level) functions
4. How the storage and comparison of biometric templates should be separated from the sensor, the server (if the biometrics application is client–server based), and the stand-alone device
5. How to securely provide remote management services to the databases in which the actual biometric templates reside in

As mentioned in this chapter, as a C-level executive, it is very important that before you make the decision to procure and acquire a biometric system, you examine the metrics that are provided to you by the vendor and challenge and/or question the validity of their claims. After all, not only you want a biometric system that will fortify your place of business but also you want to get the maximum ROI for your investment.

U.S. FEDERAL GOVERNMENT BIOMETRIC TESTING STANDARDS

To that end, there are also standards that have been developed and established at the federal level for the actual testing of the performance of biometric systems. This standard is known as ISO/IEC 19795 "Information Technology—Biometric Performance Testing and Performance." It is composed of the following parts, and each part serves as its own function in the overall biometric testing process:

1. *ISO/IEC 19795-1:2006 "Part 1: Principles and Framework"*. This part specifies how to predict and estimate various types of errors, how to recruit and use test subjects, and how to evaluate in terms of statistics the error rates received.
2. *ISO/IEC 19795-2:2007 "Part 2: Technology and Scenario Evaluation"*. This part describes the various biometric testing scenarios that can take place, such as offline and online testing, technology testing, scenario testing, and defining testing requirements.
3. *ISO/IEC TR 19795-3:2007 Technical Report "Part 3: Modality-Specific Testing"*. This part specifies the design methods for testing a singular biometric modality.
4. *ISO/IEC 19795-4:2008 "Part 4: Interoperability Performance Testing"*. This part mentions how to test biometric modalities in a heterogenous environment, how to evaluate the results of each testing scenario, and how to test various biometric sensors.
5. *ISO/IEC CD 19795-5 "Part 5: Grading Scheme for Access Control Scenario Evaluation"*. This part reviews the various testing scenarios of using biometrics in physical access entry applications.
6. *ISO/IEC CD 19795-6 "Part 6: Testing Methodologies for Operational Evaluation"*. This part involves how to get real-time, snapshot results of biometric systems that are functioning live in the field and how to determine when a biometric system is actually hacked into.
7. *ISO/IEC CD 19795-7 "Part 7: Testing of On-Card Biometric Comparison Algorithms"*. This part provides guidelines into testing smart card, biometric-based applications.

It should be noted that a C-level executive does not have to use all of the above provisions to test the biometric system he or she wants to deploy; rather, he or she can choose what part is most relevant to him or her. All of the parts that make up this testing standard are independent of each other. Also, the parts described are only the bare, minimum requirements for testing the reliability, validity, and effectiveness of a biometric system. As a C-level executive, you are expected to go above and beyond these minimums to fully ensure that you are getting the very best biometric system possible.

A REVIEW OF BIOMETRIC SENSORS

Although the algorithms are a very important component of biometric systems, there is yet another equally, if not more, important component as well, and that is the biometric sensor. The sensor has been discussed in parts of this chapter, but now is the time to explore it in much more detail and analysis.

It is the sensor that captures the actual raw image of our biological and physiological components—whether it is the eye, the face, our hand, or finger, or even the vein pattern that is present from just underneath our palms. In a sense, if there is a sensor in biometric systems, there is no way that our unique features can ever be captured.

True, a biometric system could use a camera to capture all of these information and data, but the quality of the raw images would not nearly be as good as those captured by an actual sensor. Thus, that is why biometrics technology remains primarily a contact-based one. It is because of this nature that biometrics in certain parts of the world, especially in the United States, have such a hard time of receiving wide-scale public acceptance.

For instance, even in Japan, touching a biometric sensor directly is considered to be a form of *taboo*, thus explaining why non-contact biometric technology, especially that of vein pattern recognition, is so popular in that part of the world.

Thus, what exactly is a biometric sensor? It comes in many shapes and sizes as well as technology types. For instance, with fingerprint recognition, the sensor tends to be rather small, and on a laptop device, it is a very small square. However, the sensor on a hand geometry scanner will be obviously much larger, as the entire image of the hand must be captured.

The type of sensor technology used varies with the specific biometric technology in question. For example, with fingerprint recognition, optical technology is primarily used; facial recognition utilizes special types of cameras (for lower-level applications, even a webcam can be used to capture the raw image of a face); with keystroke recognition, the keyboard is the sensor itself; and with voice recognition, microphones or telephones become the actual sensor.

Specifically, a biometric sensor can be defined as follows: "is hardware … that converts biometric input into a digital signal and conveys this information to the processing device" (*Certified Biometrics Learning System, Module 1, Biometrics Fundamentals*, © 2010 IEEE, pp. 1–28). In biometrics, there are two types of sensors: (1) dumb sensors and (2) intelligent sensors. Regardless of the technology, the bottom line is that a biometric sensor must be able to capture the raw image of a physiological or behavioral trait and the unique features from that must be captured for the biometric system to verify the identity of the individual.

This part of the chapter will now review the various types of sensors available today, in both physical- and behavioral-based biometrics technology. The first type of sensor we will examine deals with fingerprint recognition. Since it is among one of the oldest biometric technologies (along with hand geometry recognition), it should come of no surprise that it will have the most types of sensors associated with it.

Since fingerprint recognition is the most widely used biometric technology for both verification and identification scenarios, it possesses an entire gamut of technological advancements. The sensors used today in fingerprint recognition include

1. Optical scanners
2. Solid-state sensors
3. Ultrasound sensors
4. Temperature differential sensors
5. Multispectral imaging sensors
6. Touchless fingerprint sensors
   a. Reflection-based touchless finger imaging
   b. Transmission-based touchless finger imaging
   c. Dimensional touchless finger imaging

Optical Scanners

With the first type of optical sensor, this is the most widely utilized type of technology used in live-scan fingerprint image (*live scan* simply means that a finger with a pulse is required for verification and/or identification). This is used more often than any other from the list above, ranging from the stand-alone fingerprint devices to the single sign on devices used on laptops and netbooks.

In these scenarios, the finger is placed on top of what is known as a *platen* and is usually made of a glass composite. From just below this platen, light is beamed from a light-emitting diode (LED) at an upward angle, and then is reflected back from the platen onto a charge-coupled device (CCD). The CCD is simply a camera that can transpose light into electrons.

Thus, as the finger is placed onto the platen, the light captures the ridges of the fingerprint back into the CCD. At this point, the ridges appear as dark lines, and since the valleys of the fingerprint are typically not recorded or captured by the CCD, they simply appear as white spaces in between the dark lines of the ridges. Because optical sensor technology is the most widely used technology, it possesses a number of key advantages such as (1) low cost, (2) relatively high resolution, and (3) a strong ability to take into account various temperature changes (this would happen if the fingerprint device were to be used in an external environment, exposed to the weather elements).

However, along with the advantages of optical sensors, there are disadvantages as well. These include (1) high energy (or power) requirements, (2) image distortions of the ridges caused by varying conditions of the skin (such as dryness or moisture) or even the platen itself, and (3) vulnerability to spoofing.

Solid-State Sensors

The next group of sensors are known as solid-state sensors. These are proving to be very good alternatives to the traditional optical scanner and is probably the second most widely used sensor in fingerprint recognition. However, rather than using a CCD, solid-state sensors use what is known as an array of electrodes to capture the image of the ridges of the fingerprint.

A capacitance level is formed between the ridges and the electrode, and this capacitance is far greater at the ridges than the valleys of the fingerprint. Thus, the image of the fingerprint is then created from which the unique features can be extracted.

There are two types of solid sensors: (1) a sensor with enough electrodes so that the entire fingertip can be placed onto it and (2) sweep sensors, which contain very few electrodes, so the fingerprint must be literally swept from right to left (or vice versa) to capture the full image of the fingerprint. The advantages of solid-state sensor technology include (1) low cost, (2) low power consumption, and (3) its very small size. However, this type of sensor yields a very low quality, which in turn gives far fewer unique features.

Ultrasound Sensors

With ultrasound sensors, they work in very much the same way as the ultrasound machines used in hospitals and doctor's offices. From the ultrasound sensor, acoustic pulses are sent to the fingerprint, and a receiver from within the biometric device then captures the returning acoustic pulses, thus resulting in an image of the fingerprint.

A key advantage of ultrasound sensor technology is that the creation of the actual fingerprint image does not depend upon any type of visual capture; thus, as a result, it is not prone to degradation like the optical sensors.

Temperature Differential Sensors

Piezoelectric effect sensors do not use any type of light or sound to create an image of the fingerprint. Rather, this type of sensor technology relies upon pressure differences to create the image of the fingerprint. With this type of technology, electric currents are created, and since the ridges and valleys of the fingerprint are at various distances from one another, different electric intensities are created (this is also known as the *piezoelectric effect*).

With the temperature differential sensors, a pyroelectric material is utilized to generate varying levels of electric current to gauge the differences of the temperature in between the ridges and valleys of the fingertip.

Multispectral Imaging Sensors

As its name implies, multispectral fingerprint scanners create numerous images of the fingerprint using various illumination methods, such as varying wavelengths and differing levels of polarizations. These multiple images are then combined to form one overall image of the fingerprint. These multiple images actually capture features of the fingerprint from just underneath the layers of the skin.

Multispectral imaging technology was created to help compensate for a major weakness of optical sensors. That is, if there is any other object embedded in the fingerprint itself (such as a bruise, cut, or any other extraneous dirt), this can also be captured, thus creating a very distorted image of the fingerprint.

Touchless Fingerprint Sensors

The last type of sensors is a group known as the *touchless fingerprint scanners*. Meaning, there is no direct contact required by the end user, and this can be very advantageous over the direct contact sensors. For instance, these noncontact sensors can avoid the physical deformities with the finger, as well as any dirt or smudges left on the platen.

Reflection-based touchless finger imaging shines a light onto the fingerprint from different angles using just one camera. In this case, images from both the valleys and the ridges are captured, unlike the full contact sensors. A prime disadvantage of this type of sensor technology is that the end user must keep his or her finger absolutely steady, and the technology must be compliant with FBI best practices and standards.

With transmission-based touchless finger imaging, a special red light is shined through the finger. This light is aimed toward the sides of the finger. Both ridge and valley images are then subsequently captured.

With regard to three-dimensional (3-D) touchless fingerprint scanning, either parametric modeling (using cylindrical models) or nonparametric models are utilized. With the former, the fingerprint images are projected onto a cylindrical model (this actually models an actual fingerprint, and then these images are projected onto a two-dimensional [2-D] print). With the latter, mathematical algorithms are used to model where the minutiae (the unique features of the fingerprint) would be located on a rolled print, and this method can take into account the irregular shapes of the fingerprint much more so than the parametric methods.

Facial recognition is the next biometric technology that utilizes a number of sensor technologies. These include

1. CCD cameras
2. 3-D sensors, which include
   a. Active sensors with structured lighting
   b. Passive/assisted stereo sensing

CCD Cameras

It should be noted that the CCDs used in the world of facial recognition are different from those used in fingerprint recognition. CCDs used in facial recognition are actual digital cameras and can yield pictures of the face from video; the output can be black and white and even color, and can also work with other spectrums of light, not just those in the visible light range.

With CCD sensors, multiple images of the face are taken in a rapid fire succession, and then are compiled into one facial image from where the unique featured can be extracted. It is also worth mentioning here the unique features extracted from the relative distances on an individual's face (such as the relative distance of the eye from the chin). However, despite the advancements made in the CCD technology for facial recognition, it still has a number of serious flaws. For example, it is unable to discriminate against

1. The angle of the camera
2. Conditions from within the external environment (such as lighting, weather, etc.)
3. Darkness and bright light
4. Changes in the physical appearance of the face (which includes changes in facial hair, weight loss and/or gain, presence of eyeglasses, any cosmetic surgery done to the face, any coloration changes, etc.).

Another problem with CCD sensors used in facial recognition is that the images are taken as 2-D ones. For facial recognition to be truly viable, the images must be taken as 3-D ones, which can thus truly represent the face from all of the different perspectives. Therefore, the next group of sensors listed above was created to counter the disadvantages of the 2-D images.

3-D Sensors/Review of Biometric Sensors in Use

In active sensors with structured lighting, typically only one camera is used to take various pictures of the face. A special light is then used to project a special pattern (such as a geometric plane) onto the face. The differences captured from this special pattern allow for the 3-D image of the face to be constructed.

In terms of the passive/assisted stereo sensing, multiple cameras are utilized, and the 3-D images of the face are then constructed through a process known as *triangulation*. Typically, in these scenarios, two calibrated cameras are used to take pictures of the face from left to right, and vice versa. Despite the potential that 3-D sensors possess, it has a number of serious flaws, which include the following:

1. The 3-D images of the face can often contain extraneous objects from the external environment, such as *spikes* (these are just merely points of bright light reflected back into the image of the face).
2. Laser is typically used, but this can be dangerous to the end user, and laser-based 3-D images of the face possess no color or textual

information or data (meaning, there are not really enough unique features that can be extracted and used to confirm the identity of an individual).

3. The speed at which the images are captured in 3-D is much slower than the capture rates of the more traditional 2-D sensor technology.

As mentioned previously, hand geometry recognition is probably the oldest of the biometric technologies. It has proven to be one of the most viable technologies available and is used in all sorts of applications. The hand geometry scanner typically uses just one type of sensor, namely, the CCD camera, and is actually very versatile for this type of biometric technology.

For example, a set of preconfigured pegs are used to guide the individual's hand onto the platen. From this point onward, the CCD can take multiple images of the top of the hand and below the hand, and even 2-D images of the hand can be taken with special mirrors located on the sides of the hand geometry scanner.

From all of these images collected, a composite image is compiled, and the unique features from the silhouette (or shape) of the hand are then extracted. Since only the silhouette of the hand is used, the resolution of the images that are captured by the CCD is of much less quality as opposed to those images captured by the CCD in fingerprint recognition.

However, despite these advantages, the CCD technology used in hand geometry recognition suffers from a number of disadvantages:

1. It is only good for verification applications (the information and data extracted are not rich enough to be used for identification scenarios).
2. The sensor and the geometry scanner device must be placed at a level where all users can access them.
3. Any object on the hand (such as a ring or a bandage) can cause distortion of the silhouette image of the hand.

The next two biometric technologies, iris recognition and vein pattern recognition, use the same type of sensor technology—namely, what is known as near-infrared spectroscopy (NIRS). Both of these sensors are non-contact and actually yield a very strong image from which the unique features can be extracted, and high accuracy rates can be achieved as well.

With iris recognition, a monochrome digital camera is utilized and is fully automated. For example, it can detect the eye, adjust to the angle needed, shine the NIR light source into the iris of the subject in question, and take multiple images in just less than 2 s. It is the NIR light source itself that actually illuminates the iris.

Technological advancements in iris recognition have now made it possible to capture the images of the irises of people at far greater distances and also while they are on the move. The NIR light source is not noticeable or even detectable by the individuals who are using it.

With vein pattern recognition, the NIR light source can be either used via reflection or transmission. With the former, the sensor that flashes the NIR light source is located on the same side of the vein pattern recognition device. With the latter, the sensor and the NIR light source are located on different sides.

The NIR light source is flashed either onto the back of the hand or the finger to illuminate the pattern of veins. Both the hand and the finger have different absorption spectra, thus allowing for rich images to be taken.

Finally, in terms of physical biometrics, voice recognition uses the most common items for a sensor—namely, a telephone, a smartphone, or even a microphone. The end user has to recite either a set of pre-established phrases (which is known as a text-dependent voice recognition system) or any other type of verbiage (which is known as a text-independent voice recognition system).

With the former system, a highly controlled environment is needed (such as a soundproof room) to collect good quality voice samples, whereas with the latter, the above-mentioned speech media can be used very easily in any kind of environment. However, despite the availability, ease of use, and great familiarity of these sensors, the main disadvantage is that of the background noise, which is present in the external environment, as well as the quality of recordings in landline phones versus smartphones.

As was discussed in the chapter, the two behavioral-based biometric technologies are that of signature recognition and keystroke recognition. Both of these technologies use sensors that are totally unlike the physical biometric sensors just examined.

With signature recognition, the sensor is an actual writing tablet using a special type of writing stylus. After the individual is done signing, his/her signature is photographed by a special camera, the resulting image is digitized, and the special software within the signature recognition system itself then compares the images of the signature.

As will be examined in detail in Chapter 2, it is not the signature itself that is examined by the special software just described. It is the behavioral mechanisms involved when composing the signature that are examined.

However, despite the ease of use of the sensor technology, it has two distinct disadvantages: (1) it is very sensitive if different techniques are used for collecting the signature sample and (2) it cannot process very short signatures.

Keystroke recognition probably possesses the simplest of all of the biometric sensors—it is a computer keyboard itself, and all that is needed to make this sensor work is a special type of software that is developed by the keystroke recognition vendors.

Once the individual starts to type, the unique typing pattern is then recorded by the special software to confirm the identity of the person using it. The sensor (keyboard) can be used for just a one time login or it can be used throughout the entire typing session, if need be. Given that the sensor technology is already built in, keystroke recognition could prove to be viable even on smartphones.

THE DISADVANTAGES OF SENSORS

Now that we have examined the different types of sensors from both the physical and behavioral biometric perspectives, all types of sensors have a general set of disadvantages, all of which can be described as follows:

1. *Aging*: All sensor technology degrades over time, which is a direct function of how extensively that particular biometric technology is being used.
2. *Obsolescence*: The rapid advances in biometric technology and the quick pace of the research and development into newer sensor technologies can render the correct state of biometric sensors almost useless and in need of upgrading.
3. *The interoperability of the sensor*: Although it is one of the primary goals that biometric systems have the ability to work together in either a homogeneous or heterogeneous modality environment, the sensors themselves may not be interoperable with another; for example, the results from a solid-state sensor could very well be different from an optical sensor.
4. *End user training*: Each type of sensor will demand its own levels of training. For example, it may take much longer to train an end user on a fingerprint recognition device than a keystroke recognition device (of course, the level of familiarity and ease of use are key variables in this example).

TYPICAL BIOMETRIC MARKET SEGMENTS

The biometric technologies covered thus far in this chapter can serve just about any industry or market where there is a need to implement a security model or to enhance the level of an existing security infrastructure. These biometric technologies can serve just about every need, it does not take a particular device to fit one need, and they can all be used to fulfill a particular need.

The device that will be used ultimately in the end will be dictated by the needs and desires of the customer who is implementing it. Although biometrics can be used just about anywhere and even anytime, there are five key market applications for which they are used. We will provide an overview here, but more details of how each biometric technology fits into each segment will be reviewed in Chapter 2. The market applications can be defined as follows:

1. Logical access control
2. Physical access control
3. Time and attendance
4. Law enforcement
5. Surveillance

Logical Access Control

Logical access control refers to gaining access to a computer network, either at the office or via a remote connection from the employee's home, or access if he or she is on the road. Computer network access is a broad term, but for purposes of the context here, it can be defined as anything that relates to a place of business or organization. This can include accessing the corporate intranet all the way to even accessing the most confidential files and resources on a corporate server.

The security tool that is most commonly used here is the traditional username and password. Although this may have worked effectively for the last couple of decades or so, it is now definitely showing its signs of severe weaknesses. For instance, people (especially employees) write down their passwords on Post-it notes and stick them to their monitor in plain sight (this is also known as the *Post-it syndrome*). In addition, employees concoct passwords that anybody can hack into (for instance, even the word *password* is used as an actual password).

As a result, to combat these weaknesses, employers have greatly tightened their security policies, requiring employees to create passwords that are excessively complex to remember (such as requiring a capital letter, a number, or a punctuation mark at different locations of the password). All this has led to greatly increased administrative password reset costs, which can amount to as much as $300 per year per employee for any business.

To fight these escalating costs as well as employee/employer frustrations with passwords, biometrics has been called upon to replace the use of passwords totally. The use of biometrics in this regard is referred to as *single sign-on solutions* (SSOs) because with one swipe of your finger, or even just one scan of your iris, you can be logged into any computer network without having to type in any password.

Presently, fingerprint recognition is the most used biometric technology for the SSO. These biometric devices and their associated software are all very easy to install and use. The opposite is also true of logical access control: although biometrics can be used to grant access to various computer networks, files, and highly confidential resources, it can also be used to restrict and limit the access of any employee to what his or her job roles just minimally require.

Physical Access Control

Physical access entry refers to giving an individual (such as an employee) access to a secure building or even a secure room within an office building infrastructure. Traditionally, keys and badges have been used to gain access. However, the main problem is that these tools can be very easily lost, stolen, replicated, or even given to other individuals who have no business in physically being at that particular location.

Again, biometrics have been used here to replace the entire lock/key and ID badge approach. In this type of market application, a biometric device can be wired to an electronic lock strike, and once the employee has been verified by the biometric system, the door will automatically open.

The advantage to this is obvious: no more lost, stolen, or fraudulent use of keys and ID badges, and only the legitimate individual, whose identity has been 100% confirmed by the biometric system, will gain access. Any type of biometric modality will work here, but the most dominant technologies used in this type of scenario are those of hand geometry recognition and fingerprint recognition.

In a particular market application, the biometric systems can either operate in a stand-alone mode or in a networked mode. Typically, small businesses will pick the former approach, and much larger businesses will opt for the latter approach to accommodate the greater number of employees. Also, networked, physical access entry biometric systems offer succinct advantages such as

1. Greater biometric template storage capacity.
2. Larger applications (such as physical access to multiple buildings and multiple doors) can be much better served.
3. All of the biometric information and data can be stored into a central server, so an employee can access multiple doors or buildings without having the need to store their biometric templates at each and every biometric device.
4. All entrances can be wired to a central location where the administrative functions and duties can be done at once, without having the need to perform these tasks at each and every biometric device.

Time and Attendance

Businesses, at all levels of industry served, have to keep track of the hours their employees have worked so that they get paid fairly for the amount of work produced. Keeping track of the time when employees have worked can be done by numerous methods, such as using a time card, utilizing a sign in/sign out roster at the main place of entry of the business, entering the hours worked into a spreadsheet, or even using an online portal to record the clock in and clock out times.

However, using these manual approaches has proven to be a massive headache in terms of administration on at least two fronts: (1) using these manual methods of processing the hours worked and calculating payroll can be very tedious, laborious, time-consuming, and costly and (2) also, with these manual methods, fraud and forgery can very easily take place, especially with regard to what is known as *buddy punching*. This is where one employee fraudulently reports the time worked for another employee when the latter did not even physically appear for work.

Biometrics can also play an integral role in time and attendance applications as well and alleviate the above two problems. Again, any type of biometric modality can work here, but it has been hand geometry recognition and fingerprint recognition that have been traditionally utilized. However, iris recognition and even vein pattern recognition are becoming more widely used now.

Biometrics-based time and attendance systems can either operate in a stand-alone or network mode, but it is the latter that offers the most advantages. For example, a business can have central control and administrative functions from within a central server, and best of all, all of the administrative tasks associated with processing payroll and distributing paychecks can now be completely automated, thus reducing paperwork and overhead costs.

Also, using a biometric-based system, the problem of buddy punching is totally eliminated, and employees are paid fairly for the work they have produced and the hours they have worked and can even help to enforce break times, lunch, and business opening/closing time rules as well as keep tabs upon periods of employee nonproductivity.

Law Enforcement

Presently, in the United States, it is the federal government that is the biggest customer and awarder of very lucrative biometrics-based contracts to the private sector. Many of these contracts awarded are for the war on terrorism and law enforcement applications in the United States. Over the past few years, the use of biometrics by law enforcement has transcended all levels, from the national to the state to the local agencies.

As a result, law enforcement officials are quickly realizing that using the traditional methods of ink and cards for the collection and analysis of fingerprints in identifying and apprehending suspects is far becoming a thing of the past. People in law enforcement need to make split-second decisions when apprehending and questioning suspects, and biometrics can certainly give them that very exact tool they need to make those types of instantaneous decisions.

Thus, as a result, law enforcement has also become a big market application for the biometric technologies. In this market spectrum, any biometric modality will work, but it is mostly fingerprint recognition that is the most widely utilized. Other biometric technologies, such as iris recognition, facial recognition, and even vein pattern recognition, are more widely used on the current war on terrorism as well as to help identify those innocent refugees who are caught in the middle of all of this turmoil.

However, at the level of law enforcement in the United States, the primary interest in biometrics is that of identification. For example, when a police officer approaches a suspect, he or she has no way of knowing who the suspect actually is. They are just making their own best estimate from a profile content and a mug shot.

The only way to truly identify the suspect is by taking one of his/her biometrics (namely, the fingerprint) and running that image through a gargantuan database. If a match is positively made, then the police officer will know that he or she has the right person in custody. To accomplish this task, law enforcement officials throughout all of the levels rely on what is known as AFIS.

This is a huge database repository that contains the biometric templates and fingerprint images of all inmates and known criminal suspects in the United States. This repository is used primarily for identification, or in other words, conducting one too many searches. The AFIS is managed by the FBI and has actually been around for quite some time and clearly possesses distinct advantages over the traditional ink and roll and card file methods of identification.

The advantages are as follows:

1. The AFIS allows for the traditional ink and roll fingerprint images to be scanned into a special type of software and be easily converted over to biometric templates or even digitized images as specified by the AFIS file format structure.
2. Fingerprint images can be scanned directly from the finger of a suspect and be uploaded into the AFIS database in just a matter of minutes.
3. Once the fingerprints are scanned and uploaded, they can be made immediately available for transmission to other databases where they can also be stored and matched.
4. Using the AFIS, very clear images of the fingerprint can be captured, unlike the traditional ink and roll methods, where clear and distinct fingerprint images are not present.

It should be noted that the AFIS can capture fingerprint images of all ten fingers from a suspect and convert them to their respective biometric templates.

In an effort to upgrade the current AFIS, the Integrated Automated Fingerprint Identification System (IAFIS) was introduced and is currently managed by the Criminal Justice Information Services. The IAFIS offers considerable advantages over the present AFIS by containing the following features:

1. The fingerprint images (all ten digits) and other biometric information and data on some 55 million or more individuals are stored and are electronically connected to all of the law enforcement agencies across all of the 50 states.
2. Because of their electronic-based characteristics, the fingerprint images and the biometric templates are always available.

3. Results from criminals and noncriminal searches (such as conducting a background check on job applicants) can be sent to the requesting law enforcement agency in less than 24 h.
4. Latent fingerprint images collected from a crime scene are also stored into the IAFIS databases.
5. Apart from the storing of fingerprint images and biometric templates, the IAFIS databases (specifically known as the IAFIS Interstate Index) can also store the criminal history records of all criminal suspects and inmates from all over the 50 states and records can be made available to the requesting law enforcement agency within 2 h.
6. Digital criminal records and photographs of known suspects and inmates are also stored into the IAFIS databases and are available upon demand.
7. The IAFIS database also supports remote connectivity, i.e., police officers in the field can connect to the database via a WiFi-based handheld reader to conduct rapid searches on the fingerprint images, the biometric templates, and latent fingerprints.

Surveillance

The last application of biometric technology is that of surveillance. This is probably what the public fears the most, and as a result has literally impeded its growth in the United States. What scares the public most about biometrics is the covert nature in which it can be used, without having any prior knowledge that they are being watched, recorded, and subsequently analyzed by various biometric devices.

For example, facial recognition systems can be embedded very easily now into CCTV cameras, and these can be placed in public places, without anybody's knowledge of the facial recognition system being used in conjunction with the CCTV cameras. Given this covert nature of biometric technology, many citizens in the United States fear that law enforcement and especially the federal government are watching our every move. Thus, this stokes up fears of George Orwell's *Big Brother is watching*.

However, it is important to keep in mind that not all types of surveillance are associated with law enforcement. For instance, any person or any business who knows how to set up a surveillance system can implement it in a short period and potentially spy on another business competitor or even spy on customers as they shop in order to determine their buying patterns. In fact, under the umbrella of surveillance, another large market application for biometrics is that of conducting surveillance operations here and abroad on potential terrorist groups and terror suspects.

With all of this in mind, it is important to discriminate what the types of surveillance activities can actually transpire, and there are five major types:

1. Overt surveillance
2. Covert surveillance
3. Tracking individuals on watch lists
4. Tracking individuals for suspicious behavior
5. Tracking individuals for suspicious types of activities

With overt surveillance, the public, the individuals, and even the businesses in question know that they are being watched, whether it is directly disclosed or it is being perceived. The primary goal of overt surveillance is to prevent and discourage unlawful behavior.

In terms of covert surveillance, no individual or business entity has any knowledge that they are being watched and recorded. As mentioned, it is this segment of surveillance of which the public is the most afraid in terms of biometric technology usage.

With regard to watch list surveillance, the primary objective is to find an individual or group of individuals whose identity can be confirmed but their whereabouts are totally unknown. A good example of this are the so-called terror watch lists used at the major international airports in the United States. The security officials have the names of the suspects on that particular list, and if after checking a passenger's travel documents there is a match in names, then that person is further scrutinized.

In terms of the fourth type of surveillance, which is suspicious behavior surveillance, the goal here is to question individuals whose behavior tends to be very erratic, abnormal, or totally out of the norm. This can be considered to be a *macro* type of surveillance because the exact identity of the individual is not known ahead of time; instead, one is trying to filter out undesirable behavior in a group of people in a public or even private setting.

Finally, with the last type of surveillance, which is looking out for suspicious activities, an individual is on the lookout for a potential suspect whose activities are illegal and can lead to great harm to the public and businesses in general. Again, the goal here is to collect direct evidence that can be used in a court of law to convict a potential suspect before the actual activity is carried out. A good example of this is the current war on terrorism, where federal law enforcement agencies have kept tabs on potential terror suspects to ascertain if their activities are illegal or not.

The form of biometric technology that has been widely used with these types of surveillance activities has been and continues to be facial recognition. Obviously, with surveillance, you want a contactless form of biometric technology to be used—which is why facial recognition is the popular choice.

Another biometric technology that can be potentially used in surveillance applications is iris recognition. It is also non-contact, and with the recent advancements in iris recognition technology, iris images can be captured of people on the move and at even further distances.

Finally, another potential biometric technology that can be used for surveillance activities and which is still under heavy research and development is gait recognition, i.e., the actual verification and identification of an individual is based upon how they walk or their unique stride.

A key point to be made here is that these biometric technologies just mentioned do not require a cooperative subject because the most widely used surveillance activity is that of covert operations. However, despite the advantages that these biometric technologies offer to a particular type of surveillance application, all surveillance activities suffer from a number of limitations, which are as follows:

1. The suspect typically does not want his or her location to be known. For example, deciding upon where to deploy facial recognition systems to track and apprehend known suspects can thus prove to be very challenging.
2. Potential suspects can always change their physical appearances and basically spoof a biometric technology. For example, if a potential suspect knows that he or she will be in a public area for a certain amount of time, he or she can easily wear a disguise and totally subvert any facial recognition technology.
3. To avoid detection, potential suspects typically try to *hide out* in very crowded and very poorly lit places, thus even further stretching the limitations of both facial recognition and even iris recognition.

Finally, in the end, all types of security applications, whether biometric or nonbiometric, must address three levels of access controls to be truly successful:

1. What a person has (such as an ID badge)
2. What a person knows (such as a password or PIN)
3. What a person is (biometrics)

REVIEW OF CHAPTER 1

Chapter 1 provided an overview into the science and nature of biometrics. The goal was to provide a broad underpinning as to what biometrics is all about. Its primary goal was to give the background needed to comprehend the rest of the topics covered in this book.

In summary, Chapter 1 provided an introduction of what unique features are all about; an exact definition was provided of what biometrics is (as well as what it is **not**); a summary of the biometric technologies that are available (physical biometrics versus behavioral biometrics) was reviewed; a specific definition of recognition, as well as its components (verification, identification, authorization, and authentication), was examined; a discussion about biometric templates is provided (as well as the various types of biometric templates that are used today, especially those of enrollment and verification templates); the most important biometric technology and standards were explored in great detail; an overview of sensor technology and how it is used in biometrics was closely looked at; and a brief survey of the major biometric applications (such as physical access entry, time and attendance, logical access control, surveillance, and law enforcement) was done.

2

A Review of the Present and Future Biometric Technologies

THE BIOMETRIC TECHNOLOGIES OF TODAY: PHYSICAL AND BEHAVIORAL

Chapter 2 will now take a part of Chapter 1 and explore it in much more exhaustive detail, that is, biometric technologies that are available today. Each biometric technology that was presented in Chapter 1 will now be further expanded upon and reviewed.

Before we move onto our detailed discussion and analyses of the biometric technologies that are available on the market place today, it is very important to turn back to the topic of physical biometrics and behavioral biometrics. As was mentioned in Chapter 1, physical biometrics involves taking a biological or physiological snapshot of a part of our body and behavioral biometrics involves taking a picture of our unique mannerisms that make us different from everybody else.

However, we need to provide much more precise, scientific definitions to these two terms. Physical biometrics can be defined as "acquiring physical biometric samples which involves taking a measurement from subjects. This does not require any specific action by the subject. A physical biometric is based primarily upon an anatomical or physiological characteristic rather than a learned behavior" (*Certified Biometrics Learning System, Module 2, Biometrics Standards,* © 2010 IEEE, p. 2–1).

Behavioral biometrics can be defined as "Acquiring behavioral biometric samples requires subjects to be active. They must perform a specific activity in the presence of a sensor. A behavioral trait is learned and acquired over time rather than based upon biology" (*Certified Biometrics Learning System, Module 2, Biometrics Standards,* © 2010 IEEE, p. 2–1).

DIFFERENCES BETWEEN PHYSICAL BIOMETRICS AND BEHAVIORAL BIOMETRICS

As one can see, based upon these two definitions, some very subtle differences between physical biometrics and behavioral biometrics can be observed. Probably the biggest difference to be noted is the amount of activity required by the end user. With physical biometrics, no active part is needed on the part of the individual in order to collect that individual's physiological and biological unique features (whether it be the hand, the finger, the iris or retina, or even the vein patterns present from underneath the palm or the fingerprint). The end user must be cooperative in order for an effective sample to be captured (this is true for all biometric technologies).

In other words, physical biometrics can be more or less considered as *garbage in–garbage out.* The image is captured, unique features are extracted, and the individual is either verified or not. However, with behavioral biometrics, a specific function, or an active part, must be carried out (such as typing on a keyboard or signing your name), which is learned over time.

As a result, because of this learned behavior, deviations or changes in the behavioral-based biometric templates that are collected can drastically occur over time, at a level much more so than physical-based biometric templates. Hence, research and development is already underway in which a behavioral-based biometric system can literally learn and take into account such changes in the behavior and mannerisms that can occur over the lifetime of an individual.

A perfect example of this would be the use of neural network technology. With this type of technology, computer-based learning and reasoning takes place in an attempt to closely follow the actions and patterns of the human brain.

Another difference between physical biometrics and behavioral biometrics that is evident from the above definitions is the number of measurements taken to extract the unique features. In both types of technologies, multiple images or multiple samples are collected. However, in physical biometrics, only one composite image is utilized from which to extract the unique features. Thus, with physical biometrics, it is considered that only one measurement is taken.

With behavioral biometrics, multiple samples are collected, but a composite sample is not created from which to extract the unique features. Rather, statistical profiles are created from each individual sample to positively verify or identify an individual. For example, with keystroke recognition, an end user has to type multiple lines of text. Statistical profiles (using hidden Markov models [HMM] primarily) are created on the basis of each line of text typed, and from there, the unique patterns are noted, and then the individual is either fully verified or not verified at all.

WHICH ONE TO USE: PHYSICAL OR BEHAVIORAL BIOMETRICS?

A question that often gets asked is, what type of biometric system should be used? In other words, should a physical- or behavioral-based biometric system be utilized? As alluded to in Chapter 1, and which will be covered in more detail in later chapters, a big factor upon which biometric system to choose will largely depend upon the system analysis and design study, which should be conducted prior to the acquisition of a biometric system.

Another big factor is the type of security application upon which you are planning to use biometrics. For instance, as a C-level executive, are you looking at verification or identification type of scenario? In other words, are you going to confirm the identity of an individual 1:1 level or on a 1:*N* basis?

When it comes to verification scenarios, just about any biometric technology will work, whether physical or behavioral based. Probably the more robust biometric technologies to be used in this fashion would be that of fingerprint recognition, iris recognition, hand geometry recognition, and vein pattern recognition (VPR).

However, when it comes to identification scenarios, the only choice to use would be that of the physical-based biometrics. Behavioral-based biometrics are simply not robust enough to capture the identity of an individual in a large database. Also, as was discussed, behavioral-based technologies simply do not possess the capability to capture the dynamic changes in the behaviors or mannerism of an individual that occur over the course of time.

In this regard, again, physical biometrics becomes the prime choice to be used for identification types of applications. Fingerprint recognition is the best choice here, as it has been used the most and tested across very large databases.

Also, another critical factor in the choice of a physical- or behavioral-based biometric system is in the ease of the acquisition of the respective biometric templates upon the specific environment in which they operate. In other words, it is very important to choose a physical- or behavioral-based biometric that will fit into the environment in which you want it to operate.

For instance, behavioral-based biometrics will work in small office-based environments, and physical-based biometrics such as hand geometry recognition is more suited for usage in extremely harsh environments, such as factories and warehouses. Facial recognition is well suited for covert types of applications, such as tracking individuals in airport settings, or large sporting venues. On the opposite side, none of the behavioral-based biometrics are suited for large-scale environments.

Apart from all of these factors just described, there are other hurdles included as well, when choosing either a behavioral- or physical-based biometric system. These other factors include

1. The environment in which the biometric system will operate
2. The network bandwidth and data transmission needs in the transfer of the biometric templates as they traverse across the network medium
3. The size of the population and demographic specifics upon which the biometric system will be used
4. The environmental comfort of the biometric system for the end user (also known as ergonomics)
5. The ability of the biometric system to operate in conjunction with legacy security systems
6. The end user acceptance level of the biometric system that will be deployed

Remember that there is no such thing as the perfect biometric system. They all have their strengths and weaknesses as well as their technological flaws. Therefore, great thought needs to be given when planning which biometric system will best fit the needs of your business.

FINGERPRINT RECOGNITION

As mentioned in Chapter 1, fingerprint recognition has been the one biometric technology that has been around for the longest time. The use of fingerprints goes back all the way to the 1500s or so and since then has been associated with law enforcement. In fact, the fingerprint has become the de facto standard used in law enforcement.

In fact, fingerprints were found as far back as 2000 BC, as a means of a legal signature in the very primitive business transactions that took place during that time. The first true research paper that attempted to examine the unique structures of the fingerprint was published around 1694, and the very first fingerprint classification system was done in 1823 by a scientist known as Jan Purkinje. Beginning in the early 1900s, law enforcement agencies in the United States started to use fingerprints as the primary means to track down known suspects and criminals.

The following statistics just further exemplify the popularity of using fingerprints as the primary means of the verification and/or identification of individuals:

1. By 1994, the FBI databases held 810,000 fingerprints.
2. By 2003, this number had increased to more than 200 million fingerprints held in their databases.
3. By 2009, this number had escalated to 500 million plus fingerprints held in the FBI databases.

To keep up with the growing demand for the use of fingerprints, the FBI devised the Automated Fingerprint Identification System, or AFIS, to automate fingerprint-based searches across levels of law enforcement, at the national, state, and local levels. To keep up with the growing demands of this gargantuan database, the Integrated Automated Fingerprint Identification System, or IAFIS, was introduced, with enhanced features (details are given in Chapter 1). An image of a fingerprint can be seen in the following figure.

However, in the world of biometrics, the details of the fingerprint are broken down into three distinct levels:

1. Level 1: The pattern images that are present in the fingerprint
2. Level 2: The minutiae points of the fingerprint (this is from where a bulk of the unique features are actually extracted)
3. Level 3: The shapes and images of the ridges and their associated pores

The Unique Features

It is important to note that most of the biometric-based fingerprint systems only collect images at Levels 1 and 2, and it is only the most powerful fingerprint recognition systems that collect Level 3 details and are used primarily for identification purposes. Level 1–specific features include the following:

1. *Arches*: These are the ridges that just flow in one direction, without doubling back or going backward. These only comprise about 5% of the features of the fingerprint.
2. *Loops*: In this feature, the ridges go backward and go from either the left to the right or the right to the left. There are two distinct types of loops: (a) radial loops, which go downward, and (b) the ulnar loop, which goes upward on the fingerprint. These make up 65% of the features within the fingerprint.
3. *Whorls*: The ridges in the fingerprint make a circle around a core, and these comprise 30% of the features in the fingerprint.

In addition to the above features, which are collected by a fingerprint recognition system, the actual number of ridges and the way that these ridges are positioned (specifically their orientation) can also prove to be a very distinctive feature as well and help contribute to the verification and/or identification of an individual. Other distinctive features that can be extracted but are not as commonly used include the following:

1. *Prints/islands*: These are the very short ridges found on the fingerprint.
2. *Lakes*: These are the special indentations/depressions located right in the middle of the ridge.
3. *Spurs*: These are the actual crossovers from one ridge to another.

The Process of Fingerprint Recognition

Fingerprint recognition is one of those biometric technologies that work well not only for verification types of scenarios but also for identification, which is best exemplified by the gargantuan databases administered by the FBI. However, whether it is identification or verification that is being called for, fingerprint recognition follows a distinct methodology that can be broken down into the following steps:

1. The actual raw images of the fingerprint acquired through the sensor technology that is being utilized (a detailed review of various sensors can be seen in Chapter 1). At this point, a quality check is also included. This means that the raw images collected are eventually examined by the biometric system to see if there are too much extraneous data in the fingerprint image, which could interfere in the acquisition of unique data. If there is too much of an obstruction found, the fingerprint device will automatically discard that particular image and prompt the end user to place his/her finger into the platen for another raw image of the fingerprint to be collected. If the raw images are accepted, they are subsequently sent over to the processing unit, which is located within the fingerprint recognition device.
2. With the raw images that have now been accepted by the system, the unique features are then extracted and then stored as the enrollment template. If fingerprint recognition is being used by a smartphone, a smart card is then utilized to store the actual enrollment template and can even provide for some processing features for the smartphone.
3. Once the end user wishes to gain physical or logical access, he or she then has to place his or her finger onto the sensor of the fingerprint recognition system, so that the raw images and unique features can be extracted as described above, and these become the enrollment templates. The enrollment and verification templates are then compared with one another to determine the degree of similarity/nonsimilarity.
4. If the enrollment and verification templates are deemed to be close in similarity, the end user is then verified and/or dentified and is then granted physical or logical access to which he or she is seeking.

As you can imagine, as a C-level executive, the quality control checks that are put into place for the fingerprint raw images are rather extensive. The common error in thinking is that it is the enrollment and verification templates that are the most important in any type of biometric system. Although this is true to a certain extent, keep in mind that these templates are actually constructed and created from the raw images themselves.

Fingerprint Recognition Quality Control Checks

The need to collect high-quality raw images the first time is of paramount importance. The following are the quality control checks for fingerprint recognition:

1. *Resolution*: This refers to the total number of dots per inch (DPI), also known as the total pixels per inch (PPI). Most fingerprint algorithms require DPI or PPI resolutions of at least 250 to 300 DPI, and to meet the stringent requirements of the FBI, a minimum of 500 DPI is required.
2. *Area*: This is the actual size of the scanned and captured fingerprint image. The specifications set forth by the FBI mandate a minimum scanned fingerprint area of at least one square inch. Anything less than that will not produce a good-quality raw image.
3. *Frames per second*: This is the total number of raw images the fingerprint recognition device sends to the processing unit. A higher number of frames per second means a much greater tolerance for any unwanted movements of the fingerprint on the platen.
4. *The number of fingerprint pixels*: This refers to the total number of pixels in the scanned image.
5. *Dynamic range*: This refers to the possible ranges that are available for the encoding of each pixel value, and the FBI mandates at least 8 bits.
6. *Geometric accuracy*: These are the geometric differences between the enrollment and the verification templates, which are calculated via the deviations from the *x*- and *y*-axes on the fingerprint template.
7. *Image quality*: This is the variable that refers to identifying unique features in the fingerprint, such as the ridgeline patterns and various minutiae that are extracted.

Methods of Fingerprint Collection

Fingerprint images can be collected using one of two methods (or even both, if need be):

1. Offline scanning
2. Using live scan sensors

Offline scanning methods have been in existence since fingerprints have been used in law enforcement. These traditional methods involve using an inked impression of an individual's finger and then placing that impression onto a piece of paper. Subsequently, cards are utilized instead of paper to store the inked fingerprint image. These are kept in file cabinets, and when a suspect is apprehended, his/her fingerprints are then compared with the inked fingerprint image stored on the card.

Over time, as technology improved, these two-dimensional (2-D) images are then scanned into a digital format, translated into 500 DPI, to meet the FBI requirements and specifications. However, with today's fingerprint recognition systems, a law enforcement officer in the field with a wireless biometric device can scan a suspect's fingerprint and then have that image automatically uploaded into a central server, from where it can then be converted into a proper fingerprint digital image.

As one can imagine, the use of these traditional methods definitely has its fair share of flaws. For instance, it takes great skill to get a good inked impression of a fingerprint; too little ink used means that the print area of the particular finger can be missed and too much ink used could very well obscure any features that one is attempting to gather. However, the offline methods do possess one great advantage over the live-scan method, that is, a *rolled* impression of the fingerprint can be gathered, which means that a much larger image of the fingerprint can be captured versus the much smaller scan areas of the live-scan sensors.

To this degree, the question that often gets asked is, how are the latent fingerprints captured? Latent fingerprints are those that get left behind at a crime scene and are collected later in the course of an investigation. Offline scanning methods are used to collect these latent fingerprint images at a later time, and often times, chemical reagents and processes are used to collect the fingerprints.

With live-scan sensors, two methods of collecting fingerprints images are used:

1. Touch
2. Sweep

With the touch method, direct contact is required of the finger onto the biometric sensor. However, using this type of method possesses some grave disadvantages. For instance, the platen can become dirty or smudged quite easily, any fingerprints left behind on the sensor or platen by another user previously can lead to substantial errors, and in terms of dollars and cents, there is a positive correlation between the size of the scan area and the actual cost of the sensor. In addition, some sensors may not be able to capture raw images if the finger is off by more than 20° from the platen.

Unlike the touch method where only the finger is scanned, the sweep method captures images of multiple fingerprints as they are read by the fingerprint sensor, and from that point, multiple images are captured at the end of the sweep. From these multiple images, one composite image is created. The quality of the raw images that are captured are also partially dependent (as well as the other variables just described) on the accuracy of the reconstruction algorithm whose primary function is to compile the multiple, raw images into one image of the fingerprint.

The biggest disadvantage of the sweep method is that a much higher rate of error is introduced due to the variance in the sweeping rates and the angles. It is the sweep method that is primarily used by fingerprint recognition devices today, and there are three types of sensors that are available:

1. *Optical sensors*: These are the most commonly used.
2. *Solid-state sensors*: The image of the raw fingerprint is captured onto a silicon surface, and the resulting image is then translated into electrical signals.
3. *Ultrasound sensors*: This is where acoustic signals are sent toward the finger, and then, a receiver subsequently digitizes the echoes from the acoustic signals.

The Matching Algorithm

As mentioned, it is the matching algorithm that compares the enrollment template with the verification template, and to ascertain the degree of similarity or closeness between the two, a certain methodology must be followed, which is described as follows:

1. Whatever data are collected from the raw image of the fingerprint, they must have some sort of commonality with the enrollment biometric template already stored in the database. This intersection of data is known as the core, which is also the maximum curvature in a ridgeline.

2. Any extraneous objects that could possibly interfere with the unique feature extraction process must be removed before the process of verification/identification can actually occur. For example, some of these extraneous objects can be the various differences found in the size, pressure, and rotation angle of the fingerprint, and these can be normalized and removed by the matching algorithm.
3. In the final stage, the unique features collected from the raw data (which actually become the verification template) must be compared with those of the enrollment template later. At this point, this is where the matching algorithm does the bulk of its work. The actual matching algorithm can be based upon the premise of three types of correlations:
   a. *Correlation-based matching*: When two fingerprints are overlaid upon each other, or superimposed, differences at the pixel level are calculated. Although a perfect alignment of the superimposed fingerprint images is strived for, it is nearly impossible to achieve. In addition, a disadvantage with this correlation method is that these calculations can be very processing intensive, which can be a grave strain on computing resources.
   b. *Minutiae-based matching*: This is the most widely used type of matching algorithm in fingerprint recognition. With this method, it is the distances and the angles between the minutiae that are calculated and subsequently compared with one another. There is global minutiae matching as well as local minutiae matching, and the latter method focuses upon the examination of a central minutiae as well as the nearest two neighboring minutiae.
   c. *Ridge feature matching*: With this matching method, the minutiae of the fingerprint are combined with other finger-based features such as the shape and size, the number, and the position of various singularities as well as global and local textures. This technique is of great value if the raw image of the fingerprint is poor in quality, and these extra features can help compensate for that deficit.

Fingerprint Recognition: Advantages and Disadvantages

Although fingerprint recognition is one of the dominant biometric technologies available today, it is very important to take an objective view of it, from the standpoint of its advantages and disadvantages. Remember, biometric technology is just that—just another piece of a security tool. It has its fair share of flaws just like anything else. Therefore, it is critical to examine these variables as well.

By looking at the advantages and disadvantages of any biometric system, as a C-level executive, you will be in a much better position to make the best procurement decision possible. The advantages and disadvantages of fingerprint recognition as well as the other biometric technologies reviewed in this chapter will be examined from seven different perspectives:

1. *Universality*: Every human being has fingerprints, and although a majority of the world's population can technically be enrolled into a fingerprint recognition system, a small fraction cannot, and thus, a manual system must be put into place to confirm the identity of these particular individuals.
2. *Uniqueness*: The uniqueness of the fingerprint is essentially written by the DNA code, which we inherit from our parents. As a result, even identical twins are deemed to have different fingerprint structures. Interestingly enough, although the uniqueness of the fingerprint is accepted worldwide, there have been no concrete scientific studies to prove this hypothesis. Thus, the uniqueness of the fingerprint is still in theory only, unless it is supported by hardcore scientific tests, and the results prove it.
3. *Permanence*: Although the basic features of a fingerprint do not change as we get older over time, the fingerprint is still very much prone to degradation from conditions that exist in the external environment, such as cuts, contact with corrosive chemicals, etc.
4. *Collectability*: This refers to the collection of the actual raw image of a fingerprint. However, unlike other biometric technologies (especially the non-contact ones), fingerprint recognition can be affected by a number of key variables that can degrade the quality of the raw image(s) collected. These variables include moist, oily, and dirty fingers; excessive direct pressure (this is pressing the finger too hard on the sensor) as well as rotational pressure (this means that the finger rolls side to side on the sensor); dirt and oil left on the sensor; as well as residual fingerprint left by previous users (also known as latent fingerprints). Another disadvantage is that fingerprint recognition requires very close contact on the part of the end user, thus rendering this technology almost useless for high-level surveillance and identification (one too many) applications.

5. *Performance*: This is composed of two variables—accuracy and the ability to meet the specific requirements of the customer. The accuracy of a fingerprint recognition system is based upon factors such as the number of fingers that are used for both verification and identification applications, the quality and robustness of the matching algorithms utilized, and the size of the database that holds the fingerprint biometric templates. Under this performance umbrella, fingerprint recognition offers a number of key advantages:
   a. Owing to the sheer number of vendors (biometric vendors are among the largest in the biometric industry) in the marketplace translates to much lower pricing when compared with the products and solutions of other biometric vendors.
   b. Since fingerprint recognition is probably the most widely used biometric technology, this translates into easy training programs for the end user(s).
   c. The size of the biometric template (both enrollment and verification templates) is very small, on the range of 250 B to 1 kB.
   d. Owing to the advancements in fingerprint recognition technology, very small sensors have been created, thus allowing for this technology to be used on small portable devices like netbooks and smartphones.
6. *Acceptability*: This evaluation criterion indicates how much fingerprint recognition is acceptable by society as a whole. Although a majority of the world's population accept fingerprint recognition, there are also serious objections to using it, which are described as follows:
   a. Privacy rights issues: This is where the use of fingerprint biometric templates falls outside of their intended uses, which further results in the degradation of the sense of anonymity.
   b. Concerns over hygiene: All fingerprint biometric systems require direct human contact with the sensor. For example, the sensor can become contaminated with germs from other users of the system, which can be easily transmitted.
   c. There is a strong association with fingerprints and criminal activity because fingerprints have been the hallmark of law enforcement for the past number of centuries.
7. *Resistance to circumvention*: This refers to how well a fingerprint biometric system is hack-proof. That is, to what degree can a

biometric system be resistant to security threats and risks posed to it from outside third parties? Although fingerprint recognition is mostly hacker-proof, it does possess a number of vulnerabilities. Probably the biggest flaw is that of latent fingerprints that can be left behind in a sensor. As a result, these latent prints can be used to produce what is known as *dummy prints*, and these could be used to possibly fool a biometric system. However, keep in mind that while this scenario is a possibility, all fingerprint biometric systems require that a live fingerprint be read—one that is supported by temperature, electrical conductivity, and blood flow. This greatly reduces the risk of spoofing.

Market Applications of Fingerprint Recognition

As mentioned throughout in this book, fingerprint biometrics is the most widely used biometric technology; therefore, it serves a wide number of market applications, and some of the following are just examples of its enormous breadth and depth:

1. *Forensics*: The most common example is that of capturing latent fingerprints.
2. *Administration of government benefits*: The goal here is to 100% verify the identity of people who are entitled to receive government benefits and entitlements and to make sure that they are properly received.
3. *Use in financial transactions*: Fingerprint recognition is the most widely used for accessing cash at ATMs worldwide and for allowing cashless transactions. This works by associating the identity of a verified individual to that of a checking or savings account (or any other source of funds) from which money can be withdrawn, after the fingerprint is presented to the biometric system. A perfect example of this is the use of fingerprint recognition systems at grocery stores. For example, some grocery stores allow for customers to pay for their products with a mere scan of their finger. The amount is then subsequently deducted from the customer's checking account that is tied to their biometric template.
4. *Physical and logical access*: This is probably the biggest market application for fingerprint recognition systems. Physical access refers to gaining access into buildings or other secured areas from inside a physical infrastructure. Logical access refers to gaining access to

computer networks, corporate intranets, and computers and laptops themselves. In this regard, fingerprint recognition is the most widely used technology for single sign-on solutions (also known as SSOs), that is, instead of having to enter a password, all one has to present to the device is his or her fingerprint, and in less than 1 s, that individual can be logged in to his or her computer.

Case Study 1: A Biometric System for Iraqi Border Control*

There is no doubt that the recent war in Iraq has left the country with many security holes and lapses, very often now using very manual and traditional methods of tracking individuals who are entering and exiting the country. Such is the case of the situation in the Iraqi Kurdistan province of Sulaymaniyah. Before looking for a much more sophisticated solution, border control agents were manually checking in and recording the information and data of visitors entering into the province.

All of the administrative processes (especially the paperwork) were done by hand, and the average time to cover each visitor took an average of 15–20 min per visitor. Imagine if there were hundreds upon hundreds of visitors wishing to enter the province, this would cause an administrative nightmare.

Also, there were fears that these manual methods set in place were not properly confirming a visitor's true identity, and there were also grave concerns that insurgents and terrorists could very easily penetrate through the existing security infrastructure. As a result, government officials looked for a much better security solution and biometrics, especially that of fingerprint recognition, to form their own type of AFIS.

Over time, government officials subsequently installed 100 fingerprint recognition systems, with options to expand up to 400 more fingerprint recognition systems. Also, the Kurdistani government implemented a web-based border control management system to record all of the information and data pertaining to the visitors entering and exiting Kurdistan. All of these information and data are uploaded to central servers, for ease of access and storage.

The benefits of this new biometrics-based system are obvious and are as follows:

1. Increased staff productivity of up to 70%.
2. Great minimization in the processing of paperwork and the registration of visitors.

---

* Information and statistics for this case study was provided by M2SYS Technology, LLC, Atlanta, GA.

3. A 100% conversion from manual methods to automated and electronic-based tracking systems.
4. The entire biometric system is now controlled by the Kurdistani government, thus greatly reducing terrorist activities such as improvised explosive devices (IEDs) and other types of insurgent activities.

Case Study 2: Tracking Millions of Inmates and Visitors at U.S. Jails and Correctional Facilities*

The United States is known to have one of the largest prison populations in the world. In fact, there are more than 2 million American adults incarcerated in the prison systems, which include all three levels: federal, state, and local (county). In virtually all of the correctional facilities throughout these levels, the traditional method of keeping track of inmates has been a horrendous and gigantic problem, especially with prison overcrowding occurring on an almost daily basis as well as the low staff rate at these facilities who need to monitor and take care of the inmates that are incarcerated.

Some of the problems associated with the traditional methods include

1. Inmate booking
2. Medicine distribution for the inmates
3. Location and tracking of inmates
4. Releasing of unauthorized inmates due to human error
5. Swapping of ID bracelets with other inmates

As a result of these major problems, the U.S. prison system has looked at other means in which to automate current systems and to totally replace the manual methods that have caused all of these headaches and hassles. Naturally, biometrics was looked upon as the primary technology, and although not every incarceration facility has utilized biometrics, the ones that did have implemented fingerprint recognition with great success.

As a result of these biometric implementations, the immediate effects that have been realized include the following:

1. Duplication of booking entries has been eliminated.
2. ID fraud (especially that of swapping of the bracelets) has also been eliminated.
3. A fool-proof and 100% accurate method of tracking the movements of inmates has been implemented.

---

* Information and statistics for this case study was provided by M2SYS Technology, LLC, Atlanta, GA.

4. Confirming and verification of the right inmate have also been implemented.
5. Incarceration facilities can now be assured that the same person who was booked and incarcerated is actually the same person who is being released.
6. The problem of booking prisoners using the identity of another inmate (or even a close relative) has been greatly eliminated. (This was a huge problem when the old-fashioned methods of using fingerprint images stored on cards were used.)
7. The time involved in making a positive identification of a particular inmate has been greatly reduced as well.
8. The problem of releasing the wrong inmate has also been 100% curtailed because fingerprint recognition offers a means to double-check the identity of a certain inmate. When traditional methods were being used, a lot of money and time had to be spent in recapturing the right inmate and subsequently correcting the inmate records.

It should be noted also that using fingerprint recognition, many correctional facilities have also implemented the use of facial recognition and iris recognition as well, either separately or as a multimodal security solution.

HAND GEOMETRY RECOGNITION

Hand geometry recognition is even older than fingerprint recognition. This particular technology dates all the way back to the 1960s; it was patented in 1985, and by the mid-1990s, the evolution of fingerprint recognition started to take market dominance, which was previously held by hand geometry recognition. Today, there is only one vendor that develops and manufactures hand geometry scanners.

It should be noted that the hand itself does not possess any unique features; rather, it is the shape of the hand (or the geometry of the hand) that is unique among most individuals. Given this situation, hand geometry recognition is rarely used for large-scale identification purposes. Rather, it is used for verification purposes only.

In this regard, it can serve very large-scale applications for very harsh environments, such as physical access entry for warehouses, storage facilities, and factories. It is also used widely in time and attendance scenarios. In these verification scenarios, very often, a smart card is also used to store further encrypted data about the end user.

In the enrollment process, some 96 or more measurements of the hand are taken, and these include the following variables:

1. Overall shape of the hand
2. Various dimensions of the palm
3. Length and width of all ten fingers
4. Measurement between the distances of the joints of the hand
5. Various shapes of the knuckles
6. Geometrical circumference of the fingers
7. Any distinct landmark features that can be found on the hand.

Hand Geometry Recognition: Enrollment Process

To start the enrollment process with a hand geometry scanner, a combination of prisms and light-emitting diodes (LEDs) from within the scanner are used in order to help capture the raw images of the hand. The technology can capture images of the back of the hand as well as the palm. This creates a 2-D image of the hand.

To capture a three-dimensional (3-D) image of the hand, five guiding pegs are located just underneath the camera to help the end user position his/her hand properly. Although this is advantageous, this method also possesses a serious disadvantage. For instance, the images of these pegs are also captured into the raw image. This results in greatly increased processing time because the images of the pegs have to be removed.

Because of this, the extraction algorithm cannot account for variances due to hand rotation and differences in the placement of the hand. To create the enrollment template, the average of the measurements described previously is calculated, which is then converted into a binary mathematical file. This is very small—only 9 B.

A problem in the construction of the enrollment and verification templates is that the geometric features of the hand share quite a bit of resemblance or correlation with one another, which can greatly hinder the process of unique feature extraction. To help alleviate this problem, a method known as principal component analysis, or PCA, is used to produce a set of geometric features that are uncorrelated, and thus, unique features can be extracted.

The small biometric template size of hand geometry scanning gives it a rather distinct advantage. For example, hand geometry recognition works very effectively and is very interoperable as a multimodal security solution for both physical and logical access entry as well as for time and attendance applications. As a stand-alone device, a hand geometry recognition device can store more than 40,000 unique biometric templates.

Hand Geometry Recognition: Advantages and Disadvantages

As reviewed in the strengths and weaknesses of fingerprint recognition, the same seven criteria can be used to review the advantages and disadvantages of hand geometry recognition.

1. *Universality*: When compared with other biometric technologies, hand geometry recognition does very well here. This is because most end users have at least one hand that they can use to be verified. The technology has advanced to the point that it can take into account, up to a certain degree, any deformities of the hand. Also, hand geometry recognition is not affected at all by any type of skin color—it is just the shape of the hand that is captured. The technology is very easy to use and train individuals on. For example, at most, an end user needs to know how to place their hand properly onto the platen and how to swipe a smart card.
2. *Uniqueness*: Although each individual possesses a different hand shape, the unique features that are extracted do not possess the very rich information and data that the iris and retina possess.
3. *Permanence*: Although the dimensions of the hand are relatively stable throughout the lifetime of an individual, the hand is very susceptible to physical changes. This includes such factors as weight loss and weight gain, injuries, as well as various diseases.
4. *Collectability*: This is another key advantage for hand geometry recognition. Except for the variables just described in item 3, the raw images of the hand that are captured are not affected at all by things that could affect the surface of the skin, and this includes primarily grime, dirt, and scars. Thus, this key strength makes hand geometry recognition very suitable for very harsh and extreme weather environments.
5. *Performance*: Overall, hand geometry scanners are very easy to use, and its very small template size (only 9 B) makes this technology very interoperable with other types of security systems. In addition, the accuracy rate of hand geometry recognition is very high—ranked with an equal error rate (ERR) of 0.1% (see Chapter 1 for a detailed explanation of the ERR, as well as other important performance metrics). However, hand geometry scanners

are large and bulky, and this can prove to be a very serious disadvantage.

6. *Acceptability*: For the most part, hand geometry recognition is well received by the end user population and is perceived to be very noninvasive. The only major negative aspect of this type of biometrics is the hygiene of the hand geometry scanner, such as any germs or contaminants left behind by other users.
7. *Resistance to circumvention*: Even though a hand geometry scanner is very easy to use, it is very difficult to spoof. Trying to fool this kind of system would mean creating an entire 3-D physical mock-up of the hand.

As mentioned earlier, hand geometry recognition is used primarily for physical access entry and time/attendance applications. However, it has also made its way at the international airport setting and is used primarily in the U.S. Citizenship and Immigration Service Passenger Accelerated System, also known as INSPASS.

This program has been designed so that frequent flyer passengers, especially business travelers, can be enrolled into what is known as a *trusted travel program*. With this, a passenger can enter into the country of their destination with a mere scan of their hand, rather than wasting time at customs and immigration lines using the traditional paper passport. Currently, this program is in use in the United States, Canada, and Israel.

Case Study 3: Yeager Airport*

Ever since the tragic events of 9/11, security at major international airports and airlines have become of prime importance. In this regard, biometrics has become the tool of choice for the greatly enhanced security measures that have been taken and continue to take place. Biometrics is being used for all types of airport security applications, but most notably for physical access entry, logical access entry, and time and attendance.

Although it is important to screen passengers traveling through the airports, it is equally, if not more, important to screen the airline and airport employees, who pose just as much of a security threat. As a result, today, even smaller and medium-sized airports across the United States are starting to implement biometric technology.

A prime example of this is Yeager Airport, located in Charleston, West Virginia. It currently occupies a land space of 737 acres, and six

* Information and statistics for this case study was provided by Schlage Recognition Systems, Inc., Campbell, CA.

commercial airlines serve the airport as well as the surrounding communities. The need for increased security was great. For example, the airport control tower is located right in the heart of the main passenger terminal.

The airport has currently 80 employees, and the majority of them access this terminal, with the only access point being the main entrance doors. Other statistics that demonstrate the need for enhanced security are the following:

1. The main entrance doors open an average of five times per hour, round the clock.
2. The airport had initially hired a team of security professionals to guard these doors 24 h a day.
3. The cost of maintaining this security staff around the clock was a staggering $1200 per day.
4. The airport's HVAC system as well as other sensitive equipment are located in a basement near a stairwell, which leads to the control tower.

Ultimately, Schlage Recognition Systems provided the biometric security tool that would meet the need for enhanced security: hand geometry recognition scanners. This particular technology provided can record at least 31,000 points and can take some 90 or more measurements of an airport employee's hand (which include the length, the width, the thickness, and the surface) in just a matter of less than a second.

Once an airport employee has been properly verified and authenticated by hand geometry recognition scanners, only then will the control tower doors open. The prime benefit of implementing this sort of biometric technology has been the eliminated need for security personnel to watch and monitor the main doors, thus resulting in a huge cost savings for the airport.

Five hand geometry recognition scanners were initially installed, with more implementations being planned for the future. As the airport director, Nick Atkinson, summed up: "It has been the consensus since 9/11 that using biometrics in access control validation is the way to go."

VEIN PATTERN RECOGNITION

VPR is also known as *vascular pattern recognition*, or simply VPR. Of all the biometric technologies reviewed in this chapter, this technology is the most up and coming in the market place and is giving very serious competition to fingerprint recognition and, to a certain degree, even iris recognition.

VPR is considered to be what is known as an automated physical biometric. This simply means that no direct contact is required by the end user on a VPR system. Thus, this classifies VPR as a non-contact biometric technology, thus making it greatly appealing to the public and customers at large. VPR examines the unique pattern of blood vessels in either the palm or the finger (in this case, multiple fingers can be used).

A question that often gets asked at this point is why are veins chosen for this particular technology instead of arteries? The answer is quite simple, for three primary reasons:

1. Veins are much larger than the arteries.
2. Veins are much closer to the surface of the skin, whereas the arteries are located much below the skin surface.
3. Veins carry deoxygenated blood throughout the entire body.

It is the third feature of the physiology and the anatomy of the veins that form the science behind VPR. For example, to extract the unique features of the vein pattern, a near-infrared (NIR) light is exposed to either the palm or the fingertips. Hemoglobin absorbs the NIR light. Hemoglobin is a pigment in the blood stream that is heavily constituted of iron and carries oxygen in the bloodstream. The veins contain a much lesser concentration of hemoglobin than the arteries. As a result of this, the veins absorb much more NIR light, and thus, they appear much darker and robust when the raw images are captured. The following figure shows an illustration of the vein patterns that are found from the palm. One may question where the veins get their unique pattern. Just like the unique features of the iris and the fingerprint, the unique pattern of veins is created and formed from within the first 8 weeks of gestation, and hence, this unique vascular pattern is then set from the moment we are born.

Components of Vein Pattern Recognition

The actual VPR technology is composed of four major components, which are as follows:

1. A sensor with LEDs that emit the NIR light as previously described
2. A high-resolution charge-coupled device camera
3. A processing unit that extracts the unique pattern of blood vessels from either the palm or the fingertips and that also produces the enrollment and verification templates
4. The database that contains the enrollment and verification templates as well as a detailed log history of the verification and identification transactions that have transpired during a set period.

How Vein Pattern Recognition Works

With a VPR system, the end user can place either their palm or their fingertips within the scanner, and to properly position them, guides can be used. The latter part is the only contact that is required, and there is no direct contact with either the palm or the fingertip on the actual sensor itself, thus giving VPR its non-contact feature.

From this point onward, after the palm or fingertip is properly positioned, the NIR light is then flashed toward the target, and to illuminate the unique pattern of blood vessels, two techniques are used, which are described as follows:

1. *Diffused illumination*: With this method, the NIR light source and the sensor are located on the same side of the VPR. As a result, the NIR light, which is reflected back from either the palm or the fingertip, is then used to capture the raw vein pattern. The thickness of the skin is not a factor in this method.
2. *Direct illumination*: With this technique, the NIR light is flashed directly at either the palm or the fingerprint and through the target. This gives the impression of the ability to see the vein patterns right through the skin. This technique is directly affected by the thickness of the skin. The NIR light can be flashed from the top, below, or on the sides of the VPR scanner. All three have their own set of advantages and disadvantages, which are as follows:
   a. *Top lighting*: This gives the most robust raw image of the vein patterns, but these lighting units can be large and can be exposed very quickly to dirt and grime.

b. *Side lighting*: This requires extra NIR lighting power as well as much more processing power to capture the raw image of the vein pattern.
c. *Bottom lighting*: This is the cheapest option, but it is much more sensitive to the actual NIR light and for capturing the raw images of the vein pattern originating from a fingertip.

From within the VPR system itself, specialized software is used to normalize the raw images of the vein pattern that has just been captured and also removes any types of kinds of obstructions such as skin hair on the palms or on the fingertips. The extraction algorithms then extract the unique features on the vein pattern, and the enrollment template remains encrypted in the database of the VPR system.

Vein Pattern Recognition: Advantages and Disadvantages

Like fingerprint recognition and hand geometry recognition, VPR can also be rated and compared against the same seven criteria:

1. *Uniqueness*: Scientific studies have shown that even identical twins have unique vein patterns and structures.
2. *Universality*: VPR technology has an acceptance rating of very near 100% and is very versatile to use when compared with the other biometric technologies. For instance, as described, either the palm or multiple fingers can be used, and if need be, even a combination of both the palm and the fingers can be used, thus providing for a very effective multimodal biometric solution.
3. *Permanence*: Questions are still being raised if the vein pattern does indeed change in structure and physiology over the lifetime of an individual, but currently, it poses no serious issues for the purposes of verification and identification with VPR technology.
4. *Collectability*: VPR technology is very easy to use, manage, and deploy. Verification is very quick (less than 1 s), and since this is a non-contact technology, the raw images of the vein are not at all impacted by dirt on the surface of the skin, cuts, bruises, scars, or even moisture and dryness on the fingertips or the palm. The only variable that can impact the quality of the vein pattern raw image is in the thickness of the skin itself.
5. *Performance*: VPR possesses two very distinct advantages over the traditional biometric technologies:

a. The matching algorithms are not at all complex and are also very *compact* in nature; thus, the processing and storage demands toward a VPR system are quite low.
b. The sensor technology has made great advancements, and hence, it has become very small in nature, thus positioning VPR technology as a strong potential to be used for smartphone security.
c. It should be noted, however, that a VPR system can be very prone to the negative effects of ambient light or light that comes from the external environment.

6. *Acceptability*: Because of its non-contact nature, VPR technology is viewed much more as a *hygienic* form of biometric technology (especially in countries like Japan), thus giving VPR its great appeal to customers. There are also virtually no privacy rights or civil liberty issues with VPR technology because an individual's vascular pattern is basically invisible to the outside world.
7. *Resistance to circumvention*: It is very difficult to spoof a VPR system because just like the other biometric technologies, a VPR system requires a live subject, i.e., a constant blood flow in the veins is required for the raw images to be captured.

Given how well VPR ranks against these seven criteria as well as its wide public acceptance, it is an accepted fact that VPR serves a wide range of markets, even specialized markets. VPR serves both governmental and commercial applications. Examples are as follows:

1. Confirming the identity of government welfare program recipients.
2. Using VPR as an SSO solution for logical access entry, thus eliminating the use of passwords entirely.
3. VPR is also seeing heavy usage as an SSO at point of sale (POS) terminals.
4. Wide-scale usage in physical access entry scenarios (primarily for office building type of entry, VPR is not meant for the harsh environments like warehouses, factories, etc.—this is where hand geometry recognition comes into play).
5. VPR is also seeing a huge increase in the usage at the major international airports as a way of quickly and efficiently confirming the identity of passengers via the use of their e-passport.
6. VPR is an excellent choice for time/attendance applications for all types of workforces.
7. VPR is even used at car rental places, mini storage units, as well as condominiums and dormitories in universities and colleges.

Case Study 4: Yarco Company*

Business owners, regardless of what industry they are in and the customers they serve, all face the administrative nightmares of the human resource function. Many of the headaches stem from having to calculate the total number of hours each employee has worked for the week, and from that point, computing the appropriate payroll and taxes that are to be distributed using old-fashioned paper-and-pencil methods.

Herein lies another big problem: buddy punching. This occurs when an employee takes an unexcused absence from work and has another employee clock in their hours for them and thus getting paid for work that was not at all performed. Such are the problems that were faced by Yarco.

This particular entity is a multifaceted real estate firm, which has more than 12,000 apartment complexes located across 11 different states and has a market capitalization in excess of $600 million. This business also has a construction division with revenues over $90 million.

Many employees of Yarco were manually inputting their time worked onto traditional paper timesheets. The payroll department would then have to collect all of these timesheets from over 100+ Yarco locations and then manually enter all of these data onto Excel spreadsheets. This entire process would take at least one weekend to complete.

On top of that, the timesheets were faxed over to the payroll department and were very often hard to read. To make matters even worse, there was no formal review process put into place to approve all of these timesheets. As a result, all of the timesheets submitted by employees were assumed to be legitimate no matter what, thus further exacerbating the problem of buddy punching.

To remedy this problem, and to save the company from the burgeoning costs of payroll processing, Yarco implemented VPR into their business. The company has a VPR system installed at each of their apartment complexes, for a total of more than 100 VPR systems. They also implemented a central server that houses all of the employee and payroll information and data.

The benefits that were realized by Yarco as a result of implementing VPR technology has been immense:

1. A 90% increase in efficiency in payroll administration using VPR versus using the traditional methods.

* Information and statistics for this case study was provided by M2SYS Technology, LLC, Atlanta, GA.

2. Total elimination of manually entering the time worked for each employee onto spreadsheets.
3. A 50% reduction in the administrative overhead devoted to payroll processes.
4. Total wipe out of buddy punching.
5. A centralized system with a web-based software application that allows staff and management to review the payroll records quickly and easily.
6. All that is needed on the part of the employee is their unique vein pattern—everything else is calculated very easily and within minutes.

PALM PRINT RECOGNITION

Another biometric technology that relates to that of hand geometry recognition is that of palm print recognition. There is a great deal of confusion between VPR and palm print recognition in that the two are very similar. However, the two are actually quite different from each other. Although one can measure the unique pattern of blood vessels from the palm itself, palm print recognition is just that: the recognition of the shape or the geometry of the palm.

As a technology, it is probably the least used in the world of biometrics. However, just like the use of fingerprints, the images of the palm have been around for quite a long time. For example, dating all the way back to the British Civil Service, palm prints were used to confirm a worker's identity when it was time to receive their pay. The very first palm recognition system that was fully automated was manufactured and introduced in Hungary back in 1994, for use by law enforcement officials.

In terms of forensics purposes, palm prints play a very crucial role as latent evidence—in fact, experts claim that up to 30% of major crime scenes can contain latent palm prints, which are subsequently used later in a court of law. One of the biggest obstacles with palm recognition is that the palm is much larger than that of any inked or sensing surfaces.

This means that the image of the palm has to be captured from two entirely separate images. These images must then be overlapped with another and then literally *stitched* together to form the entire image of the palm. From this stitched image, the features of the palm can be very easily seen. These include the creases, ridge patterns, and minutiae.

How Palm Print Recognition Works

The science and technology behind palm print recognition is virtually the same as that of fingerprint recognition. The raw images of the palm are extracted, the templates are created, and the verification/identification takes place via the comparison of the enrollment templates versus the verification templates.

Palm print technology can use optical, thermal, or even ultrasound sensors. The question of how unique the palm print is often arises. Although there have been no concrete scientific studies to prove the validity of any type of uniqueness to be found in the palm, palm prints are created and formed in the womb and remain fairly stable after the individual is born.

However, keep in mind that the palm is very much prone to threats from the external environment, such as dirt, scars, injuries, grime, and degradation of its features over the lifetime of an individual.

Palm Print Recognition: Advantages and Disadvantages

Because palm print recognition is not widely used, it can be rated against only four of the performance criterion:

1. *Collectability*: Because the palm consists of a much larger surface, there is a higher probability of collecting more unique features than of other biometric techniques (such as fingerprint recognition, where the scan surface is obviously much smaller).
2. *Performance*: Palm print recognition works very well when it is used as a part of a multimodal security solution such as hand geometry recognition.
3. *Acceptability*: Palm print recognition can be prone to the social issues surrounding biometrics because palm prints can be associated with certain types of diseases, which include Down syndrome, fetal alcohol syndrome, Aarskog syndrome, and Cohen syndrome.
4. *Resistance to circumvention*: One advantage of palm print recognition in this aspect is that it is much more difficult to spoof. For example, there are many features of the palm that have to be replicated and a much larger *dummy* palm has to be fashioned.

FACIAL RECOGNITION

Facial recognition is one of those biometric technologies that most people can associate with. For example, we all have a face, and just like the fingerprint, the face has been used to verify and identify criminals and wanted suspects as well as terrorists. Probably the best examples of facial identification are the photos at the post office and those facial images on the wanted portion of the major law enforcement websites.

However, unlike the other biometric technologies being used today, facial recognition is very prone to privacy rights issues and claims of civil liberty violations. The primary reason for this is that facial recognition can be used very covertly, without the knowledge or the consent of the individuals whom the system is trying to track down.

Also, facial recognition does have its fair share of technological flaws as well. For example, if a facial recognition system were to capture the image of an individual who is grossly overweight and then capture another image of the same person who went through massive weight loss, the facial recognition system would not be able to make a positive match. In other words, the system can be very easily spoofed in these aspects.

However, it is not just weight loss that can trick a facial recognition system but other things such as the presence and the subsequent removal of facial hair, aging, and the presence and absence of other objects on the face, such as hats, sunglasses, and switching from contact lenses to eyeglasses and vice versa. However, one of the key advantages to facial recognition is that it can be used for both verification and identification scenarios and for heavy-duty usage as well.

For instance, facial recognition is used quite frequently in the e-Passport infrastructures of many nations around the world and is also used for large-scale identification applications at the major international airports, especially in hunting down those suspects on the terrorist watch lists.

Facial Recognition: How It Works

Facial recognition technology relies upon the physical features of the face (see the following figure) that are determined by genetics. Also, this technology can be deployed either as a fully automated system or as a semi-automated system. With the latter, no human interaction is needed and all of the verification and identification decisions are made by the technology itself. With the latter, human intervention to a certain degree is required, and this is actually the preferred method for deploying a facial recognition system.

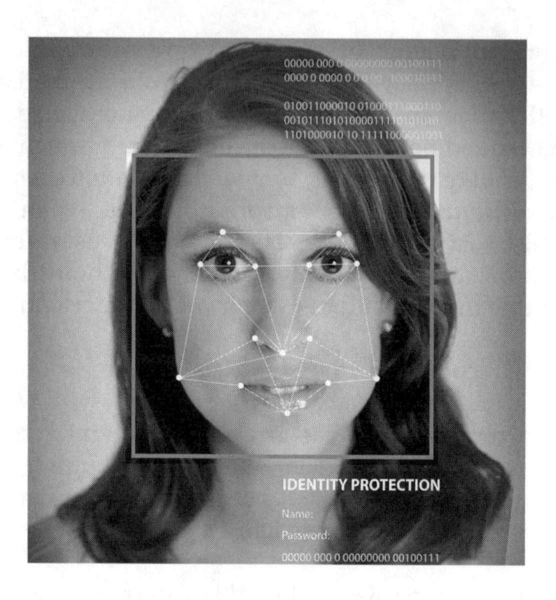

Given some of the serious obstacles it still faces, it is always better to err on the side of caution and have an actual human be involved and render the verification or identification decision.

Facial recognition systems of today focus upon those parts of the face that are not as easily prone to the hurdles just described. These regions of the face include the following:

1. Ridges between the eyebrows
2. Cheekbones
3. Mouth edges
4. Distance between the eyes
5. Width of the nose
6. Contour and the profile of the jawline
7. Chin

The methodology to capture the raw images of the face is much different when compared with the other biometric technologies. Although facial recognition is a non-contact technology, the image capture processes are much more complex, and more cooperation is required on the part of the end user. To start the process of raw image collection, the individual must first either stand before a camera, or unknowingly, have their face captured with covert surveillance methods, such as using a closed-circuit television (CCTV) camera system (with the technology that is available today, facial recognition can literally be implanted in a CCTV).

Once the raw images are collected by the camera, the data are then either aligned or normalized to help refine the raw images at a much more granular level. The refinement techniques involved include adjusting the face to be in the middle of the pictures that have been taken, and adjusting the size and the angle of the face so that the best unique features can be extracted and later converted over to the appropriate verification and enrollment templates.

All of these are done via mathematical algorithms. As mentioned previously, facial recognition is countered by a number of major obstacles, but even more so at the raw image acquisition phase. These include a lack of subtle differentiation between the faces and other obstructive variables in the external environment, various different facial expressions and poses in subsequent raw image captures, and the capture of a landmark orienting feature such as the eyes.

To help compensate for these obstacles, much research and development has been done in the area of what is known as 3-D imaging. In this technique, a shape is formed and created, and using an existing 2-D image, various features are created; the result is a model that can be applied to any 3-D surface and can also be used to help compensate for the abovementioned differences.

However, it should be noted that these types of 3-D facial recognition systems are not widely deployed in commercial applications yet because this technique is still considered to be under the research and development phases. Right now, it is primarily 2-D facial recognition systems that are used on the commercial market. 3-D facial recognition systems are only used as a complement to the 2-D ones, in which higher imaging requirements are dictated, and the capture environment is much more challenging.

Defining the Effectiveness of a Facial Recognition System

According to the International Committee for Information Technology Standards, also known as INCITS, the raw image of a face must meet certain stringent requirements to guarantee its effectiveness and reliability in a facial recognition system. These requirements are as follows:

1. The facial raw image must include an entire composite of the head, the individual must possess a full head of hair, and the raw image should capture the neck and shoulders as well.
2. The roll, pitch, and yaw of the facial raw images collected must possess a variance of at least ±5° of rotation.

3. Only plain and diffused lighting should be used to capture the facial raw images.
4. For verification and/or identification to take place, no shadows whatsoever should be present in the raw images collected.

If a 3-D facial recognition system is used, the following properties must be observed:

1. Stereo imaging must utilize at least two cameras, which are mounted at a fixed distance.
2. If structured lighting is used, the facial recognition system flashes a defined, structured pattern at the face, which is used to capture and compute depth.
3. Laser scanners are the most robust form of sensing but are very costly to implement as well as very slow; it takes as long as 30 s or even more to capture and process the raw image of a face.
4. Hybrid sensors do exist and can use both the stereo imaging and structured lighting techniques.

The entire process of facial recognition starts with the location of the actual image of a face within a set frame. The presence of the actual face can be sensed or detected from various cues or triggers, such as skin color, any type of head rotation, the presence of the facial or even head shape, as well as the detection and presence of both sets of eyes in the face.

Some of the challenges involved in locating the face in the frame include identifying the differentiation between the tonality of the skin color and the background and the various shapes of the face (depending of course on the angle in which the raw image is actually presented to a facial recognition system), or even multiple images of faces may be captured into a single frame, especially if the facial recognition system has been used in a covert fashion in a very large crowd.

Techniques of Facial Recognition

To help alleviate these obstacles and provide a solution in which a single facial image can be detected in just one frame, various techniques have been developed and applied to facial recognition. These techniques fall under two categories:

1. Appearance based
2. Model based

With appearance-based facial recognition techniques, a face can be represented in several object views; it is based on one image only, and no 3-D models are ever utilized. The specific methodologies here include PCA and linear discriminant analysis (LDA). Model-based facial recognition techniques construct and create a 3-D model of the human face, and after that point onward, the facial variations can be captured and computed. The specific methodology here includes elastic bunch graph mapping. All of these techniques will now be discussed in detail.

PCA, which is a linear-based technique, dates all the way back to 1988, when it was first used for facial recognition. This technique primarily uses what is known as *eigenfaces*. Simply put, eigenfaces are just merely 2-D spectral facial images that are composed of grayscale features.

There are literally hundreds of eigenfaces that can be stored in the database of a facial recognition system. When facial images are collected by the system, this library of eigenfaces is placed over the raw images and are superimposed over one another. At this point, the level of variances between the eigenfaces and the raw images is then subsequently computed and averaged and then different weights are assigned.

The end result is a one-dimensional (1-D) image of the face, which is then processed by the facial recognition system. In terms of mathematics, PCA is merely a linear transformation in which the facial raw images get converted over into a geometrical coordinate system. Imagine a quadrant-based system. With the PCA technique, the data set with the greatest variance lies upon the first coordinate of the quadrant system (this is also termed the first PCA), the next data set with the second largest variance falls onto the second coordinate, and so on, until the 1-D face is created.

The biggest disadvantage with this technique is that it requires a full frontal image, and as a result, a full image of the face is required. Thus, any changes in any facial feature require a full recalculation of the entire eigenface process. However, a refined approach has been developed, thus greatly reducing the calculating and processing time required.

With LDA, a linear-based analysis, the face is projected onto a vector space, with the primary objective of speeding up the verification and identification processes by cutting down drastically on the total number of features that need to be matched.

The mathematics behind LDA is to calculate the variations that occur between a single raw data point from a single raw data record. Based from these calculations, the linear relationships are then extrapolated and formulated. One of the advantages of the LDA technique is that it can actually take into account the lighting differences and the various types of facial expressions that can occur, but a full-face image is required.

After the linear relationship is drawn from the variance calculations, the pixel values are captured and statistically plotted. The result is a computed raw image that is just simply a linear relationship of the various pixel values. This raw image is called a fisher face. Despite the advantages, a major drawback of the LDA technique is that it requires a large database.

Elastic bunch graph matching (EBGM), a model-based technique, looks at the nonlinear mathematical relationships of the face, which includes such factors as lighting differences and the differences in the facial poses and expressions. This technique uses a similar technique that is used in iris recognition, known as Gabor wavelet mathematics.

With the EBGM technique, a facial map is created. The facial image on the map is just a sequencing of graphs, with various nodes located at the landmark features of the face, which include the eyes, edges of the lips, tips of the nose, etc. These edge features become 2-D distance vectors, and during the identification and verification processes, various Gabor mathematical filters are used to measure and calculate the variances of each node on the facial image.

Then, Gabor mathematical wavelets are used to capture up to five spatial frequencies and up to eight different facial orientations. Although the EBGM technique does not at all require a full facial image, the main drawback with this technique is that the landmarks of the facial map must be marked extremely accurately, with great precision.

Facial Recognition: Advantages and Disadvantages

Facial recognition systems can also be evaluated against the same set of criteria. In this regard, there is a primary difference between this and the other biometric technologies: that is, while the face may not offer the most unique information and data like the iris and the retina, facial recognition can be very scalable and like fingerprint recognition and hand geometry recognition, facial recognition can fit into a wide variety of application environments.

The evaluation of facial recognition can be broken down as follows:

1. *Universality*: Unlike all of the other biometric technologies, every individual possesses a face (no matter what the condition of the face is actually in), so at least theoretically, it is possible for all end users to be enrolled into a facial recognition system.

2. *Uniqueness*: As mentioned, facial recognition is not distinctly unique at all; even members of the same family, as well as identical twins, can genetically share the same types of facial features (when it comes to the DNA code, it is the facial features from which we inherit the most resembling characteristics).
3. *Permanence*: Given the strong effect of weight gain and weight loss (including the voluntary changes in appearance) as well as the aging process we all experience, permanence of the face is a huge problem. In other words, the face is not at all stable over time and can possess a large amount of variance. As a result, end users may have to constantly be re-enrolled in a facial recognition system time after time, thus wasting critical resources and processing power.
4. *Collectability*: The collection of unique features can be very difficult because of the vast differences in the environment that can occur during the image acquisition phase. This includes the differences in lighting, lighting angles, and the distances at which the raw images are captured, and also the extraneous variables such as sunglasses, contact lenses, eyeglasses, and other types of facial clothing.
5. *Performance*: In this regard, facial recognition has both positive and negative aspects, which are as follows:
   a. *Accuracy*: Facial recognition, according to recent research, has a false acceptance rate (FAR) of 0.001 and a false rejection rate (FRR) of 0.001.
   b. *Backward compatibility*: Any type of 2-D photograph can be added quite easily in the database of the facial recognition system and subsequently utilized for identification and verification.
   c. *Lack of standardization*: Many facial recognition systems are in existence, but there is a severe lack of standards among the interoperability of these systems.
   d. *Template size*: The facial recognition biometric template can be very large, up to 3000 B, and as a result, this can greatly increase the storage requirements as well as choke off the processing system of the facial recognition system.
   e. *Degradation*: The constant compression and decompression and recycling of the images can cause serious degradation to the facial images that are stored in the database over a period.

6. *Acceptability*: In a broad sense, facial recognition can be widely accepted. However, when it is used for surveillance purposes, it is not at all accepted, because people believe that it is a violation of privacy rights and civil liberties. Also, some cultures, such as the Islamic culture, prohibit the use of facial recognition systems where women are required to wear head scarves and hide their faces.
7. *Resistance to circumvention*: Facial recognition systems can be very easily spoofed and tricked by 2-D facial images.

Applications of Facial Recognition

Given the covert nature and the ability to deploy facial recognition easily into other nonbiometric systems and technologies, it is no wonder that it has a wide range of market applications. Probably the biggest application for facial recognition has been that for the e-passport and, to a certain degree, the national ID card system for those nations that have adopted it.

For example, the International Civil Aviation Organization (also known as the ICAO) has made facial recognition templates the de facto standard for machine-readable travel documents. Also, along with iris recognition, facial recognition is being used quite heavily at the major international airports, primarily for the covert surveillance purposes, in an effort to scan for individuals on the terrorist watch lists.

In addition, facial recognition can be used very covertly at venues where large crowds gather, such as sporting events and concerts, hence the lack of public acceptance. Facial recognition systems can also be used in conjunction with CCTV cameras and strategically placed in large metropolitan areas. For example, the city of London is a perfect example of this. At just about every street corner, there is a facial recognition/CCTV camera system deployed.

Because of this vast network of security over London, their police were able to quickly apprehend and bring to justice the terrorist suspects who were involved in the train bombings that took place. Another popular application for facial recognition is for border protection and control, which is especially widely used in European countries.

Facial recognition is also heavily used in conducting real-time market research studies. For instance, it can be used to gauge a potential customer's reaction to certain advertising stimuli by merely recording that individual's facial movements. Casinos are also a highly favored venue for facial recognition, as a means to identify and verify the welcomed guests versus the unwelcomed guests.

Facial recognition has also been used in both physical access entry as well as time and attendance scenarios, but nowhere near to the degree that hand geometry recognition and fingerprint recognition are currently being used in these types of market applications.

Case Study 5: Razko Security Adds Face Recognition Technology to Video Surveillance Systems*

Older-generation video technology includes such tools as multiplex/time lapse cameras, digital video recorders (DVRs), as well as network video recorders (NVRs). Although these traditional technologies have proved their worth over time, the expense of maintaining them over time has proliferated, for many businesses, no matter how large or how small.

A very strong, inherent disadvantage of this traditional security technology is that these are primarily sophisticated archiving mechanisms. Meaning, potential criminals and suspects can only be seen after the particular footage is recorded; thus they cannot be identified in real time and apprehended immediately. In other words, the potential criminal or suspect could very well leave the crime scene far behind and elude law enforcement officials long after the video footage has been carefully examined by security personnel.

Also, depending upon the size of the business, it is a huge expense to maintain a staff of security guards to identify the literally thousands of individuals that enter and exit the premises on a daily basis. However, with the addition of facial recognition technology to current CCTV technology, the identification of potential suspects and criminals can now occur in real time, and thus, apprehension can occur very quickly on scene, just seconds after the crime has occurred.

Although it sounds complex, the idea behind all of this is quite simple. A facial recognition database of all potential criminals and suspects is created and implemented into the CCTV technology. As the particular individual walks past the CCTV cameras, the facial images are then instantaneously compared with the database, and thus, a potential, positive match can be made.

Through the facial recognition technology developed by Cognitec, an alarm is sounded in real time once a positive match is made between the facial recognition databases and the CCTV camera

* Information and statistics for this case study was provided by Cognitec Systems, GmBH, Dresden, Germany.

footage. This is perfectly illustrated by Razko Security. Its president, Ted Eliraz, was approached by Mike Kavanaugh, a Canadian Tire business owner, based out of Oakville, Ontario.

Their business requirements were as follows:

1. To have multiple CCTV cameras with facial recognition technology implemented at various places of business
2. To have a central facial recognition database for ease of administration
3. The most important—a reasonable cost

With the facial recognition technology provided by Cognitec, the streaming video is displayed in real time, and when a positive match is made, the face of the potential criminal or suspect is then displayed, along with any appropriate law enforcement details and information about the individual in question. When installing this biometrics-based system, some big hurdles had to be overcome, in particular

1. Determining the exact locations of the CCTV/facial recognition cameras at the places of business
2. The various lighting conditions encountered from the external environment

After conducting a thorough systems analysis and design, the cameras were located at the main entrance, the exteriors, and the glass doors at the various Canadian Tire locations. In the end, Canadian Tire had this technology implemented on more locations throughout Canada, and the results proved to be very successful.

For instance, within the first 6 months of operation, four suspects were apprehended, and with further implementation of this technology at other Canadian Tire dealership locations, the facial recognition databases of potential suspects and criminals have grown, thus resulting in a much higher apprehension and capture rate.

THE EYE: THE IRIS AND THE RETINA

Of all of the biometric technologies available today, it is the eye, and its subcomponents, especially the retina and the iris, which possesses the most unique information and data in which to identify and verify an individual. As of now, it is the retina that is deemed to be the most unique biometric of all (of course, after the use of DNA, which will be much further explored later in this chapter). The iris is deemed to be the next most unique biometric, right after the retina.

The Iris

The popularity of iris recognition dates all the way back to the mid-1990s, when the first mathematical algorithms were developed and created by Dr. John Daugmann of the University of Cambridge. At the time, only one iris recognition vendor existed, known as Iridian Technologies. For well over a decade, this company owned the intellectual property as well as the mathematical algorithms and the intellectual property involved with iris recognition at the time. As a result, a very strong monopoly was held, and customers could only utilize just one type of iris recognition product or solution.

The system, although considered to be very advanced at the time, was quite expensive and bulky to deploy and maintain. However, eventually, just a couple of years ago, Iridian Technologies was bought out by a much larger biometrics vendor, and at the same time, the patents of the original algorithms and technology also expired, thus breaking the monopoly that was held for such a long time.

As a result, there has been a huge explosion of iris recognition just in the past couple of years, offering many new types of products and solutions for the customer. The advantage of this is the customer now has a much greater depth and breadth of picking and even customizing his or her own iris recognition solution that will fit his or her exact needs and requirements, unlike before, where people were basically forced to buy just one product or solution.

In fact, iris recognition has developed so quickly that now images of the iris can be captured at much greater distances as well as when people are in movement. Previously, an end user had to stand directly in front of the iris recognition camera at a very close proximity. Even the unique pattern of the blood vessels can also be scanned, and this will be examined after the section on retinal recognition.

The Physiological Structure of the Iris

The iris lies between the pupil and the white of the eye, which is known as the sclera. The color of the iris varies from individual to individual, but there is a commonality to the colors, and these include green, blue, brown, and, rarely, hazel; in the most extreme cases, a combination of these colors can be seen in the iris. The color of the iris is primarily determined by the DNA code inherited from our parents.

The unique patterns of the iris start to form when the human embryo is conceived, usually during the third month of fetal gestation. The phenotype of the iris is shaped and formed in a process known as chaotic morphogenesis, and the unique structures of the iris are completely formed during the first 2 years of child development.

The primary purpose of the iris is to control the diameter and the size of the pupil. The pupil is that part of the eye that allows for light to enter into the eye, which in turn reaches the retina, which is located at the back of the eye. The amount of light that can enter the pupil is a direct function of how much it can expand and contract, which is governed by the muscles of the iris. The iris is primarily composed of two layers: a fibrovascular tissue known as (1) the stroma, which, in turn, is connected to a group of muscles known as (2) the sphincter muscles.

The sphincter muscles are responsible for the contraction of the pupil, and another group of muscles known as the dilator muscles governs the expansion of the pupil. When you look at your iris in the mirror, you will notice a radiating pattern. This pattern is known as the trabecular meshwork. When NIR light is flashed onto the iris, many unique features can be observed. These features include ridges, folds, freckles, furrows, arches, crypts, coronas, as well as other patterns that appear in various discernible fashions.

Finally, the collaretta of the iris is the thickest region, which gives the iris its two distinct regions, known as the pupillary zone (this forms the boundary of the pupil) and the ciliary zone (which fills up the rest of the iris). Other unique features can also be seen in the collaretta region. The iris is deemed to be one of the most unique structures of human physiology, and in fact, each individual has a different iris structure in both eyes. In fact, scientific studies have shown that even identical twins have different iris structures. An image of the actual iris can be seen in the following figure.

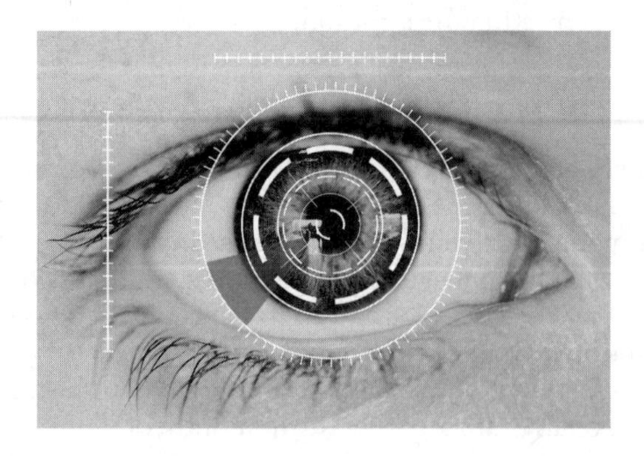

Iris Recognition: How It Works

The idea of using the iris to confirm the identity of an individual dates all the way back to 1936, when Frank Burch, an ophthalmologist, first proposed the idea. This idea was then patented in 1987, and by the mid-1990s, Dr. John Daugmann of the University of Cambridge developed the first mathematical algorithms for the application of iris recognition technology. Traditional iris recognition technology requires that the end user stand no more than 10 inches away from the camera.

With the NIR light shined into the iris, various grayscale images are then captured and then compiled into one primary composite photograph. Special software then removes any obstructions from the iris, which can include portions of the pupil, eyelashes, eyelids, and any resulting glare from the iris camera.

From this composite image, the unique features of the iris (as described before) are then *zoned off* into hundreds of phasors (also known as vectors), whose measurements and amplitude level are then extracted (using Gabor wavelet mathematics) and then subsequently converted into a binary mathematical file, which is not greater than 500 B. Because of this very small template size, verification of an individual can occur in less than 1 s.

From within the traditional iris recognition methods, this mathematical file then becomes the actual iris biometric template, which is also known as the IrisCode. However, to positively verify or identify an individual from the database, these iris-based enrollment and verification templates (the IrisCode) must be first compared with one another. To accomplish this task, the IrisCodes are compared against one another byte by byte, looking for any dissimilarities among the string of binary digits.

In other words, to what percentage do the zeroes and the ones in the iris-based enrollment and verification templates match up and do not match up against one another? This answer is found using a technique known as *Hamming distances,* which is even used in iris recognition algorithms of today.

After these distances are measured, tests of statistical independence are then carried out, using high-level Boolean mathematics (such as exclusive OR operators [XOR] and masked operators). Finally, if the test of statistical independence passed, the individual is then positively verified or identified, but if the tests of statistical independence failed, then the person is **not** positively verified or identified.

The Market Applications of Iris Recognition

As a result of these technological breakthroughs in iris recognition, it now cuts across all realms of market applications. In the past decade, because of the monopolistic grip it held, iris recognition served only a very limited number of market applications, and because of that, customer acceptance, as well as the believability in the viability of this technology was very low. In addition, at that time, iris recognition technology was very expensive (about $3000–$5000), thus providing for a much bigger obstacle in its adoption rate. Today, the price has come down substantially, thus helping in boosting its acceptance by customers very quickly. Because of this, iris recognition has now become a dominant player in those market applications that were once traditionally held by hand geometry recognition and fingerprint recognition. These market applications include the following:

1. *Logical access entry*: This deals with gaining access to servers and corporate networks and intranets.
2. *Physical access entry*: This addresses issues of gaining access to secure areas in a facility or place of business.
3. *Time and attendance*: This deals with accurate time reporting and actual hours worked by employees.
4. *Critical infrastructures*: This includes nuclear power plants, oil refineries, large-scale military installations, as well as government facilities.
5. *Airports*: This includes everything from confirming passengers' identity as they make their way through the security checkpoints to reading their e-passports before they disembark at the country of their destination.
6. *Seaports*: This involves all maritime activities, which include securing marine terminals and *high-consequence* facilities such as oil and gas storage and chemical, intermodal, and port operations.
7. *Military checkpoints*: This includes securing, with iris recognition, such areas as military bases, air force bases, and naval bases.

Iris Recognition: Advantages and Disadvantages

Iris recognition, when evaluated against the seven criteria, actually fares very well in terms of both acceptance and applicability for the markets it serves.

1. *Universality*: In a general sense, everybody has at least one eye that can be scanned and from which the unique features are extracted; thus, literally, the technology can be used anywhere around the world. Even in the unfortunate chance that an individual is blind in one or both eyes, technically, the iris can still be scanned, even though it can be much more difficult for the iris recognition system to scan the actual image of the iris.
2. *Uniqueness*: As mentioned, the iris, along with the retina, is one of the richest biometrics in terms of unique information and data. Even among identical twins, the irises are also unique. For example, Dr. John Daugmann, who originally developed the iris recognition algorithms, calculated that the statistical probability of the iris being identical among twins is 1 in 10^{78}.
3. *Permanence*: One of the biggest advantages of the iris is that it is very stable, and in fact, it hardly changes over the lifetime of an individual. Thus, in this regard, iris recognition does not possess the same flaws as facial recognition does. In addition, the iris is considered to be an *internal* organ; thus it is not prone to the harsh conditions of the external environment, unlike the face, hand, or fingers. However, it should be noted that any direct, physical injury to the iris, or even a laser-based vision corrective surgery, will render the iris useless.
4. *Collectability*: When compared with all of the other biometric technologies, the images of the iris are extremely easy to collect, both in terms of camera capture and the software analysis of the iris. In fact, the images of the iris can still be captured even when an individual is still wearing their eyeglasses or contact lenses. The only disadvantage of iris recognition in this particular regard is that the image of the iris can be very much prone to blurring by the capture camera, and any excessive head tilts on the part of the individual can cause the iris image to be greatly distorted.
5. *Performance*: Given the very small size of the iris recognition template, this technology appears to be extremely fast as well as extremely accurate. For example, it possesses an FAR level of 0.0001 and an FRR at a remarkable level of 0.0. As a result of these metrics, iris recognition works very well for identification-based applications, as this is best exemplified by the iris recognition deployments in both Iraq and Afghanistan, where the technology is being used everyday to identify hundreds of refugees in those particular countries.

6. *Acceptability*: Because iris recognition is non-contact technology, overall, the acceptance has been pretty high. However, initially, there can be some unease and fear of having one's iris scanned. It should be noted that the NIR light used in iris recognition is not visible to humans, and there has been no physical harm cited as of yet to the eye as a result of using iris recognition.
7. *Resistance to circumvention*: Iris recognition technology is very difficult to spoof because the system can discriminate a live iris from a fake iris by carefully examining the dilation and constriction of the pupil, and the use of encryption to further protect the iris recognition templates greatly prevents against reverse engineering.

Case Study 6: The Afghan Girl—Sharbat Gula*

Probably the best example of using these traditional iris recognition techniques was in trying to confirm the identity of Sharbat Gula, also known as the "Afghan Girl." The famous picture of the striking, green eyes of Gula goes back almost 30 years, to 1985. She is originally from Afghanistan. During the time of the Soviet Union invasion of Afghanistan, she and her family fled into the Nasir Bagh refugee camp, located in Peshawar, Pakistan.

During that time, a world famous photographer from *National Geographic Magazine*, Steve McCurry, came to this region to capture the plight of the Afghanistan refugees on pictures. One of the refugees he met and took pictures of was Gula, when she was only 13 years old. McCurry did not know her name or anything else about her family. Of the many pictures he took of her, one stood out in particular—her piercing, green eyes.

During the course of the next 17 years, this picture became world famous, appearing in books, magazines, newspapers, posters, and other forms of media. Gula knew nothing of her fame until she met McCurry for the second time in January 2002. In January 2002, McCurry returned to the same region in a final attempt to locate Gula.

McCurry and his team searched through numerous villages and came across various leads that proved to be false. Finally, the break came when an individual came forward and claimed that Gula was a next-door neighbor many years ago. After several days of making this claim, this same individual brought back the brother of Gula, who had the

* Information and statistics for this case study was provided by BiometricNews.Net, Inc., Chicago, IL.

same color of eyes. From that moment onward, McCurry and his team felt that they had located the family of Gula.

Because of her culture, Gula was not allowed to meet other men. However, a female producer of National Geographic was allowed to initially meet with and take photographs of her. Finally, after a series of *negotiations* with her family, McCurry was able to see Gula. After asking some questions and comparing the world famous photograph to photographs just taken of her, McCurry felt he had finally found the Afghan girl—17 years later.

However, various tests had to be conducted to make sure that Gula was truly the Afghan girl. Two sophisticated tests were utilized: facial recognition techniques developed by forensic examiners at the FBI and iris recognition techniques developed by Dr. John Daugman and Iridian Technologies.

The pictures taken in 1985 were compared with the pictures taken in 2002, in both tests. The facial recognition techniques confirmed her identity; however, the ultimate test came down to iris recognition because of its reliability, as stated before in this article. However, the scientists at Iridian Technologies had to overcome a number of obstacles.

First, the pictures taken by National Geographic of Gula were not taken by iris recognition cameras; rather they were taken with other types of cameras. As a result, the scientists had to eliminate the effects of the light reflections produced by these cameras and had to make various modifications to the image quality of these photographs.

The pictures of Gula were then eventually scanned into a digital format. The next obstacle then was that iris recognition works by examining scans from live subjects, not static photographs—another major obstacle. After making a series of adjustments to the iris recognition software, the scientists concluded that Sharbat Gula was positively the Afghan girl.

The Retina

The next component of the eye to be examined is the retina. Very often, there is a lot of confusion between the retina and the iris. As discussed in length, the iris is located on the front of the eye, but the retina is located at the back of the eye. Unlike the iris, the retina is actually a group of blood vessels that lead into the optic disc, and from there, visual information is transmitted to the brain for processing via the optic nerve, which lies in between the optic disc and the brain.

The scanning of the retina is also known as retinal recognition and is currently the most unique biometric of all because of the richness of the data it possesses, from which the unique features can be extracted.

Retinal recognition technology was first conceived of in the mid-1970s, but the first retinal recognition biometric device did not come out onto the marketplace until the early 1980s. Although retinal recognition does possess some very distinct advantages over the other biometric technologies, this technology has extremely limited market applications.

The primary reasons are the high cost of retinal recognition devices (which can be as high as $5000 per device), its bulkiness, and the fact that it is very invasive. Although retinal recognition technology is a non-contact technology, the end user has to sit very close to the scanner and literally place his or her eye onto the scanning device, thus making it very cumbersome and very uncomfortable for the end user. An image of the retina is shown in the following figure.

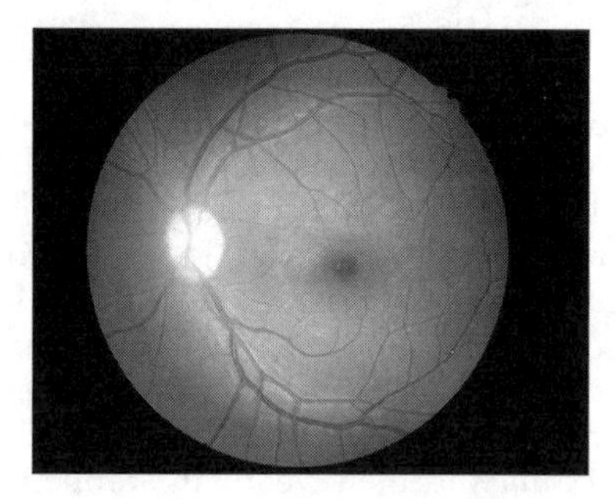

The overall side view of the eye, which includes both the retina and the iris, can be seen in the following figure.

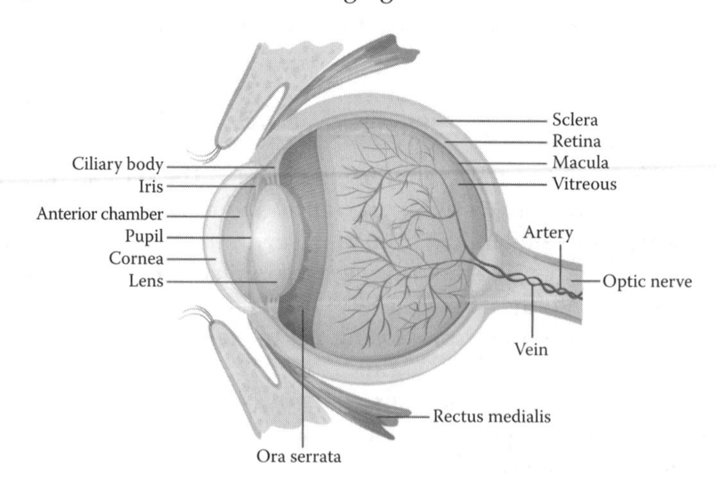

The Physiology of the Retina

It is said that the retina "is to the eye as film is to a camera" (Robert *Buzz* Hill, *Retina Identification,* from *Biometrics: Personal Identification in Networked Society,* by Anil Jain, Ruud Bolle, and Sharath Pankati, p. 124). The retina consists of multiple layers of sensory tissue and millions of photoreceptors whose function is to transform light rays into electrical impulses. These impulses subsequently travel to the brain via the optic nerve, where they are converted to images. Two distinct types of photoreceptors exist within the retina: the rods and the cones.

Although the cones (of which each eye contains approximately 6 million) help us see different colors, the rods (which number 125 million per eye) facilitate night and peripheral vision. It is the blood vessel pattern in the retina that forms the foundation for retinal recognition as a science and technology.

Because of its position within the eye, the retina is not exposed to the external environment. As a biometric, it is therefore very stable. It is from here that information is transmitted to and received from the brain. The circle in the diagram indicates the area that is typically captured by a retinal scanning device. It contains a unique pattern of blood vessels.

There are two famous studies that have confirmed the uniqueness of the blood vessel pattern found in the retina. The first was published by Dr. Carleton Simon and Dr. Isodore Goldstein in 1935 and describes how every retina contains a unique blood vessel pattern. In a later paper, they even suggest using photographs of these patterns as a means of identification. The second study was conducted in the 1950s by Dr. Paul Tower. He discovered that, even among identical twins, the blood vessel patterns of the retina are unique and different.

The first company to become involved in the research, development, and manufacture of retinal scanning devices was EyeDentify Inc. The company was established in 1976 and its first retina capturing devices were known as *fundus cameras.* Although intended for use by ophthalmologists, modified versions of the camera were used to obtain retina images. The device had several shortcomings, however. First, the equipment was considered very expensive and difficult to operate.

Second, the light used to illuminate the retina was considered too bright and too discomforting for the user. Further research and development yielded the first true prototype scanning device, which was unveiled in 1981. The device used infrared light to illuminate the blood vessel pattern of the retina. The advantage of infrared light is that the blood vessel pattern in the retina can *absorb* such light much faster than other parts of the eye tissue.

The reflected light is subsequently captured by the scanning device for processing. In addition to a scanner, several algorithms were developed for the extraction of unique features. Further research and development gave birth to the first true retinal scanning device to reach the market: the EyeDentification System 7.5. The device utilized a complex system of scanning optics, mirrors, and targeting systems to capture the blood vessel pattern of the retina.

Ongoing development resulted in devices with much simpler designs. Later scanners consisted of integrated retinal scanning optics, which sharply reduced manufacturing costs (compared with the EyeDentification System 7.5).

The last retinal scanner to be manufactured by EyeDentify was the ICAM 2001, a device capable of storing up to 3000 templates and 3300 transactions. The product was eventually withdrawn from the market because of its price as well as user concerns.

The Process of Retinal Recognition

The overall retinal scanning process may be broken down into three subprocesses:

1. *Image/signal acquisition and processing*: This subprocess involves capturing an image of the retina and converting it to a digital format.
2. *Matching*: A computer system is used to verify and identify the user (as is the case with the other biometric technologies reviewed in previous articles).
3. *Representation*: The unique features of the retina are presented as a template.

The process for enrolling and verifying/identifying a retinal scan is the same as the process applied to other biometric technologies (acquisition and processing of images, unique feature extraction, template creation).

The image acquisition and processing phase is the most complicated. The speed and ease with which this subprocess may be completed largely depends on user cooperation. To obtain a scan, the user must position his/her eye very close to the lens.

To safeguard the quality of the captured image, the user must also remain perfectly still. Moreover, glasses must be removed to avoid signal interference (after all, lenses are designed to reflect). On looking into the scanner, the user sees a green light against a white background. Once the scanner is activated, the green light moves in a complete circle (360°). The blood vessel pattern of the retina is captured during this process.

Generally speaking, three to five images are captured at this stage. Depending on the level of user cooperation, the capturing phase can take as long as 1 min. This is a very long time compared with other biometric techniques. The next stage involves data extraction. One very considerable advantage of retinal recognition becomes evident at this stage.

As genetic factors do not dictate the pattern of the blood vessels, the retina contains a diversity of unique features. This allows up to 400 unique data points to be obtained from the retina. For other biometrics, such as fingerprints, only 30–40 data points (the minutiae) are available.

During the third and final stage of the process, the unique retina pattern is converted to an enrollment template. At only 96 B, the retina template is considered one of the smallest biometric templates.

As is the case with other biometric technologies, the performance of the retinal scanning device may be affected by a number of variables, which could prevent an accurate scan from being captured. Poor quality scans may be attributable to the following:

1. Lack of cooperation on the part of the user. As indicated, the user must remain very still throughout the entire process, especially when the image is being acquired. Any movement can seriously affect lens alignment.
2. The distance between the eye and the lens is incorrect and/or fluctuates. For a high-quality scan to be captured, the user must place his or her eye near the lens. In this sense, iris scanning technology is much more user friendly; a quality scan can be captured at a distance of up to 3 feet from the lens.
3. A dirty lens on the retinal scanning device. This will obviously interfere with the scanning process.
4. Other types of light interference from an external source.
5. The size of the user's pupil. A small pupil reduces the amount of light that travels to (and from) the retina. This problem is exacerbated if the pupil constricts because of bright lighting conditions, which can result in a higher FRR.

Retinal Recognition: Advantages and Disadvantages

All biometric technologies are rated against a set of performance standards. As far as retinal recognition is concerned, there are two performance standards: the FRR and the ability to verify rate (AVR).

As mentioned in Chapter 1, the FRR describes the probability of a legitimate user being denied authorization by the retinal scanning system. Retinal recognition is most affected by the FRR. This is because the factors described above have a tangible impact on the quality of the retinal scan, causing a legitimate user to be rejected. Also, the AVR describes the probability of an entire user group being verified on a given day. For retinal recognition, the relevant percentage has been as low as 85%. This is primarily attributable to user-related concerns and the need to place one's eye in very close proximity to the scanner lens.

Like all other biometric technologies, retinal recognition has its own unique strengths and weaknesses. The strengths may be summed up as follows:

1. The blood vessel pattern of the retina rarely changes during a person's life (unless he or she is afflicted by an eye disease such as glaucoma, cataracts, etc.).
2. The size of the actual template is only 96 B, which is very small by any standards. In turn, verification and identification processing times are much shorter than they are for larger files.
3. The rich, unique structure of the blood vessel pattern of the retina allows up to 400 data points to be created.
4. As the retina is located inside the eye, it is not exposed to (threats posed by) the external environment. For other biometrics, such as fingerprints, hand geometry, etc., the opposite holds true.

The most relevant weaknesses of retinal recognition are as follows:

1. The public perceives retinal scanning to be a health threat; some people believe that a retinal scan damages the eye.
2. User unease about the need to position the eye in such close proximity of the scanner lens.
3. User motivation: Of all biometric technologies, successful retinal scanning demands the highest level of user motivation and patience.
4. Retinal scanning technology cannot accommodate people wearing glasses (which must be removed prior to scanning).
5. At this stage, retinal scanning devices are very expensive to procure and implement.

As retinal recognition systems are user invasive as well as expensive to install and maintain, retinal recognition has not been as widely deployed as other biometric technologies (particularly fingerprint recognition, hand). To date, retinal recognition has primarily been used in combination with access control systems at high-security facilities.

This includes military installations, nuclear facilities, and laboratories. One of the best-documented applications involves the State of Illinois, which used retinal recognition to reduce welfare fraud by identifying welfare recipients (thus preventing multiple benefit payments). This project also made use of fingerprint recognition.

Retinal recognition can also be compared with the seven criteria the other technologies have been compared with. However, while retinal recognition does possess some very significant advantages, overall, it does not fare as well (meaning the negatives far outweigh the positives), which seriously limits and curtails its adoption rate in the market place.

1. *Universality*: Virtually everybody, unless they have some extreme form of blindness, possesses a retina; thus at least on a theoretical level, people can have their retina scanned.
2. *Uniqueness*: As described, the retina is very unique, even among identical twins. This is probably its greatest strength.
3. *Permanence*: Unlike the other physiological components of our bodies, the biological fundamentals of the retina hardly change in the lifetime of the individual. However, it should be noted that if anything affects the iris, then the retina can be subject to degradation as well. Also, it should be noted that the retina is prone to degradation as well via diabetes, glaucoma, high blood pressure, and even heart disease.
4. *Collectability*: The scan area of the retina is very small, and the end user must place their eye in a very user-invasive eye receptacle; also, he or she must remove his or her contact lenses or eyeglasses for a quality scan to be captured. Because of this very constrained environment, the end user must remain cooperative, and the systems administrator must make sure that the retina is scanned in an appropriate time frame; if not, the failure to enroll rate (FER) will be high.
5. *Performance*: As a flip side to the user invasiveness involved, retinal recognition has extremely high levels of accuracy; in fact, it is claimed that the error rate is as low as 1 in 1 million.

6. *Acceptability*: Retinal recognition works best with those end users who are absolutely required to use it to perform their required job functions.
7. *Resistance to circumvention*: It is almost impossible to spoof a retinal recognition system, and a live retina is required for either verification or identification to take place.

Because of the high expense and user invasiveness involved, retinal recognition is only used for ultrahigh-security applications, such as for the government and the military. As a result, there is a high level of preference given to iris recognition over retinal recognition.

VOICE RECOGNITION

Voice recognition is a biometric technology whose research and development dates all the way back to World War II. For example, at that time, spectrographs showed that there are variations in the intensity of various sounds in a person's voice and at different frequency levels.

This propelled the idea of perhaps using voice recognition to confirm the identity of a particular individual. The research and development into voice recognition continued well into the 1960s, and the voice spectrographs that were used at the time started to utilize statistical modeling as a means of biometric template creation, rather than using the traditional approaches.

This continued trend would allow for the evolution of automated voice recognition tools to come into play. In fact, the first known voice recognition systems was called the *forensic automatic speaker recognition*, or FASR.

In today's biometric world, voice recognition can be considered to be both a behavioral- and a physical-based biometric. This is so because the acoustic properties of a particular person's voice are a direct function of the shape of the individual's mouth, as well as the length and the quality of the vocal cords (the physical component). However, at the same time, the behavioral data of an individual's voice are present in the template as well, and this includes such variables as the pitch, volume, and rhythm of the voice.

Voice Recognition: How It Works

The first step in voice recognition is for an individual to produce an actual voice sample. Voice production is a facet of life that we take for granted, and the actual process is complicated. The production of sound originates at the vocal cords. In between the vocal cords is a gap. When we attempt to communicate, the muscles that control the vocal cords contract.

As a result, the gap narrows, and as we exhale, this breath passes through the gap, which creates sound. The unique patterns of an individual's voice are then produced by the vocal tract. The vocal tract consists of the laryngeal pharynx, oral pharynx, oral cavity, nasal pharynx, and the nasal cavity. It is these unique patterns created by the vocal tract that are used by voice recognition systems.

Although people may sound alike to the human ear, everybody, to some degree, has a different or unique annunciation in their speech. To ensure a good-quality voice sample, the individual usually recites some sort of text, which can either be a verbal phrase, a series of numbers, or even repeating a passage of text, which is usually prompted by the voice recognition system. The individual usually has to repeat this a number of times.

The most common devices used to capture an individual's voice samples are computer microphones, cell (mobile) phones, and landline-based telephones. As a result, a key advantage of voice recognition is that it can leverage existing telephony technology, with minimal disruption to an entity's business processes. In terms of noise disruption, computer microphones and cell phones create the most, whereas landline-based telephones create the least.

Factors Affecting Voice Recognition

However, it is very important to note that the medium used to enroll the end user into the voice recognition system is the same type of medium used for later uses. For example, if a smartphone was used to create the enrollment template, then the same smartphone should be used in subsequent verification transactions in the voice recognition system.

There are also other factors that can affect the quality of voice samples other than the noise disruptions created by telephony devices. Examples are factors such as mispronounced verbal phrases, different media used for enrollment and verification (using a landline telephone for the enrollment process, but then using a cell phone for the verification process), as well as the emotional and physical conditions of the individual.

Finally, the voice samples are converted from an analog format to a digital format for processing. These raw voice data types are inputted into a spectrograph, which is a pure visualization of the acoustic properties of the individual's voice.

The next steps are unique feature extraction and creation of the template. The extraction algorithms look for unique patterns in the individual's voice samples. To create the template, a *model* of the voice is created. There are two statistical techniques that are primarily utilized when formulating the voice recognition biometric templates:

1. *HMMs*: This is a statistical model that is used with text-dependent systems. This type of model displays such variables as the changes and the fluctuations of the voice over a certain period, which is a direct function of the pitch, duration, dynamics, and quality of the person's speaking voice.
2. *Gaussian mixture models, or GMMs*: This is a state-mapping model, in which various types of vector states are created that represent the unique sound characteristics of the particular individual. Unlike HMMs, the GMMs are devoted exclusively for use by text-independent systems.

However, just like other biometric technologies, there are other external factors that can impact the quality of the voice recorded (this recorded voice can also be considered the *raw image*) and subsequently affect the quality of the enrollment and verification templates created. These factors are as follows:

1. Any type of misspoken, misunderstood, or misread text phrases the end user is supposed to recite
2. The emotional state of the individual when they are reciting their specific text or phrase
3. Poor room sound acoustics
4. Media mismatch (for instance, using different telephones or microphones for both the enrollment and verification phases)
5. The physical state of the individual (for instance, if they are sick, suffering from any other kind or type of ailment, etc.)
6. The age of the individual (for example, the vocal tract changes as we get older in age).

Voice Recognition: Advantages and Disadvantages

When voice recognition is compared against the seven major criteria, it has a mixed result:

1. *Universality*: Voice recognition is not at all language dependent. This is probably its biggest strength. As long as an individual can speak, theoretically, they can be easily enrolled into a voice recognition system.
2. *Uniqueness*: Unlike some of the other biometrics, such as the iris and the retina, the voice does not possess as many unique features that results in the lack of rich information and data.
3. *Permanence*: Over the lifespan of an individual, and with age, the voice changes for many reasons, which include age, fatigue, any type of disease of the vocal cords, as well as the emotional state and any medication the individual might be under.
4. *Collectability*: This is probably the biggest disadvantage of voice recognition. For example, any type of variability in the medium that is used to collect the raw voice sample can greatly affect or skew the voice recognition biometric templates. Therefore, it is of upmost importance to ensure that the same type of collection medium is used for both the enrollment and verification stages. There should be no interchangeability involved whatsoever.
5. *Performance*: Because of the variableness involved when collecting the raw voice samples, it is difficult to put a gauge on how well a voice recognition system can actually perform. Also, the template size can be very large, in the range of 1500 to 3000 B.
6. *Acceptability*: This is probably one of the strongest advantages of voice recognition. The technology is actually very nonintrusive and can be deployed in a manner that is very covert to the end user.
7. *Resistance to circumvention*: When compared with other biometric technologies, voice recognition, to a certain degree, can be easily spoofed by one end user mimicking the voice acoustics of another end user. This is in large part due to the lack of unique features in the voice itself.

The Market Applications of Voice Recognition

The market applications of voice recognition have been much more limited when compared with the other biometric technologies of hand geometry recognition, fingerprint recognition, and even iris recognition. A part of the reason for this is that there is a lack of vendors actually developing voice recognition solutions.

However, on the whole, voice recognition is starting to get some serious traction, as businesses and governments worldwide have started to realize its ease of deployment and the other strong advantages it offers to the market place. Probably one of the biggest market applications voice recognition is used for is in the financial world.

For example, many of the smaller- to medium-sized brokerage institutions offer voice recognition to their customers as a means of quick verification. Rather than wasting a customer's time by inputting their PIN and Social Security Numbers, a customer can be identified very quickly by the use of voice recognition. What would take normally minutes to authenticate a customer with the traditional means of security is now literally shaved down to just mere seconds.

As a result, financial transactions for the customer can occur at a very quick pace. Also, voice recognition can be used on smartphones as a means of verification instead of having to enter a numerical-based password. Voice recognition is also currently being used in correctional facilities (to monitor the telephone privileges of the inmates), the railroad system, border protection and control, as well as certain types of physical and logical access entries.

This section of the book now deals with behavioral biometrics. With this in mind, the unique features to be extracted are all based upon the variations in each person's own mannerisms, upon which they can be verified and/or identified. We first start with signature recognition.

SIGNATURE RECOGNITION

For the longest time, signatures have been associated with law enforcement and identity, much in the same way the fingerprint has been. However, unlike the other biometrics, the use of a signature actually creates a legally binding contract or agreement. As a result, the interest in the use of signature recognition has generated much more interest than just as a tool for confirming the identity of a particular individual.

In other words, the use of a signature can do both, proliferated by the legislation enacted by the Clinton Administration, which made signatures legally binding even in e-commerce transactions. The most common form of signature with which we are most familiar is the type when we sign our name at the point-of-sale terminal at our local stores.

For instance, when you sign or swipe your credit card through the point-of-sale system, you are prompted to sign your name. The image of your signature then gets transmitted to the sensor pad and, from there, to the credit card database where specialized software then matches the signature that is stored in the central databases to what has just been presented.

If the match is authentic, the transaction is then subsequently approved, and the customer can then procure their products. However, there is often confusion between this type of signature and a *digital signature,* and there is a major difference. For example, digital signatures are not based upon the actual, physical signature itself. Rather, it is based upon encrypted bit strings that are attached to electronic documents, which can be used to authenticate and approve all types of electronic transactions and to legally approve and bind *paperless documents* and records.

The Differences between a Signature and Signature Recognition

In this same regard, there is also confusion between the actual, physical signature just described and the signature recognition itself. With signature recognition, it is very important to note that it is not the image of the actual physical signature that is stored and compared in the signature recognition database.

Rather, it is the behavioral patterns inherent to the process of signing that are stored and compared. These include the changes in timing, pressure, and speed in which the signature is composed. Although it is comparatively straightforward to duplicate the visual appearance of a signature, it is very difficult to duplicate behavioral characteristics. This process is known as *dynamic signature recognition.*

Signature Recognition: How It Works

It should be noted that signature recognition is primarily used for verification purposes only. Although, in the most theoretical sense, it could potentially be used for identification applications, in the real world, it would prove to be very impractical because of the following reasons. First, the hand can be greatly affected by genetic factors and various types of physical features (such as ailments, and the aging process of the hand we all experience at some point in time or another). Second, the signature is very dynamic and can change very quickly over time, whether the individual has intentions or not to alter his or her own signature. Third, unlike the other biometric technologies, there is virtually no permanence or long-term stability (unlike the iris or the retina) associated with signature recognition because of its ever-changing nature.

The very first signature recognition devices utilized static variables such as the height, spacing, slope, as well as the various characteristics in terms of the shaping of the letters, which are found in the signature. By the mid-1970s, signature recognition became much more dynamic in the sense that various spatial, pressure, and temporal variables were taken into consideration. These variables included such factors as the downward pressure that was applied to the pen, the level of pressure at which the pen itself is gripped at, and the angle at which the pen is held while the individual signs his or her name, and even the time it takes for the signature to be completed.

Today's signature recognition devices can now collect and analyze such variables as speed, acceleration, pauses, and the changes in pressure in which the individual signs his or her name on the special writing tablet. Neural network technology can also be incorporated with signature recognition, which can literally learn the ever so slightest changes and variations in the way an individual signs his or her name over a pre-established period and make the necessary changes to the database.

Signature recognition technology involves the use of a pen and a special writing tablet, which are connected to a local or central computer for processing and verification. To acquire the signature data during the enrollment process, an individual is required to sign his or her name several times on the writing tablet. It should be noted that the robustness of the signature recognition enrollment template is a direct function of the quality of the writing tablet.

A high-quality tablet will capture all of the behavioral variables (timing, pressure, and speed). In contrast, a low-spec tablet may not be able to capture all these variables. There are several constraints to the data acquisition phase. First, a signature cannot be too long or too short. If a signature is too long, too much behavioral data will be presented. As a result, it will be difficult for the signature recognition system to identify consistent and unique data points.

If a signature is too short, insufficient data will be captured, giving rise to a higher FAR. Second, the individual must complete the overall enrollment and verification process in the same type of environment and under the same conditions.

For example, if the individual stands during enrollment but sits down during verification, the enrollment and verification templates may vary substantially (this is attributable to the amount of support given to the arm). Once the data acquisition phase has been completed, the signature recognition system extracts unique features from the behavioral characteristics, which include the time needed to place a signature, the pressure applied by the pen to the writing tablet, the speed with which the signature is placed, the overall size of the signature, and the quantity as well as direction of the signature strokes.

With signature recognition templates, different values or *weights* are assigned to each unique feature. These templates are therefore as small as 3 kB. One of the biggest challenges in signature recognition is the constant variability in the signatures themselves. This is primarily because an individual never signs his or her signature in the same fashion on any two given successive times.

For example, the writing slope can switch tangentially left to right (and vice versa), and up and down (and also vice versa), the exact pressure put on the pen can change greatly each and every time the individual has to submit a verification template, and even a reflective light on the surface of the signature recognition capture device can indirectly cause variances in the speed as well as the timing of the signature.

Signature Recognition: Advantages and Disadvantages

The most significant benefit of signature recognition is that it is highly resistant to impostors. For example, although it is quite easy to forge a signature, it is very difficult to *mimic* the behavioral patterns inherent to the process of signing. Signature recognition is well suited to high-value transactions. For example, signature recognition could be used to positively verify the business representatives involved in a transaction before any classified documents are opened and signed.

Second, compared with the other biometric technologies, signature recognition is deemed to be noninvasive. As a result, the technology could be widely accepted by users. We have all used our signature to authorize transactions, which does not in any way impede on our sense of privacy (or privacy-related rights).

Third, there is always a concern among users of biometric systems that templates may be stolen if the system is compromised and that stolen templates cannot be replaced and/or changed. This may be true for physiological biometrics (it would, for example, be very difficult to change the structure of your fingerprint or iris). Signature recognition allows the behavioral dynamics of the way you place your signature to be changed, thus easing user concerns.

On the downside, signature recognition is prone to higher error rates, particularly when the behavioral characteristics of signatures are mutually inconsistent. In addition, users may have difficulties getting used to signature tablets, which could also result in higher error rates.

Signature recognition can also be compared against the seven criteria, with the end results being that the number of advantages it offers balances out for the most part, with its disadvantages:

1. *Universality*: Probably its biggest strength, signature recognition can be used in almost any type of language, ranging from English to the most obscure languages ever heard.
2. *Uniqueness*: Obviously, the more characteristics a signature has, the more unique it becomes. As mentioned, when the variables of speed, velocity, timing, and pressure are introduced, dynamic signatures can virtually never be lost or stolen.
3. *Permanence*: This is probably one of the greatest weaknesses of signature recognition. An individual's signature can vary greatly over just a span of a few minutes by such variables as fatigue, illness, stress, or distraction. These variables in turn can also affect the ability of the individual to actually grip the writing pen used to create the dynamic signature, and exert the downward pressure that is needed.
4. *Collectability*: The ability to collect raw signature samples is a direct correlation of the quality of the signature recognition system itself (meaning, a lower-grade signature system will make it much more difficult to collect a robust signature sample).
5. *Performance*: It has been recognized that the best performing signature recognition devices are about 96% accurate. However, the better the signature recognition device, the higher the cost. A strong advantage here is that signature recognition requires no end user training.
6. *Acceptability*: In general, when compared with other biometric technologies, most people are very comfortable with providing their signature primarily because of its ease of use and non–privacy rights issues. It is also important to note that the theft of personal data is much likely lesser with dynamic recognition technology versus graphic recognition technology (such as storing a fingerprint or an iris).
7. *Resistance to circumvention*: This is one of signature recognition's other biggest advantages—a signature that is dynamically produced is almost impossible to forge.

In summary, of all biometric technologies, whether physical or behavioral, signature recognition offers the most potential in terms of adaptability and implementation. This holds true from a number of perspectives.

First, there is its ease of use. The user simply places his/her signature as he/she would normally do. There is, therefore, no need for end user training, as is the case with other types of biometric technology, including facial recognition and retinal recognition. The training needed for the person operating the signature recognition device is also minimal.

Second, the implementation cost of signature recognition is low. The system consists of a special pen, a tablet, and software (which, in the case of retailers, could be installed in a POS terminal). Costs are minimal compared with those of much more complicated systems, such as a retinal scanning device.

Third, a signature recognition system can easily be embedded in an organization's prevailing security processes, without excessively disrupting or affecting existing operations. To give an example, no major wiring or installation is needed (as is the case with a hand geometry scanner or fingerprint scanner). Signature recognition could therefore prove a highly valuable tool for multimodal security solutions.

KEYSTROKE RECOGNITION

Keystroke recognition is the next behavioral biometric technology to be examined. As its name implies, keystroke recognition relies upon the unique way in which we type onto a computer keyboard. In today's mobile world and global society, typing is something in which we take for granted everyday.

Whatever way we type onto our keyboard on our office or home computer, or even our smartphone, there is a distinct way in which we type. This uniqueness comes most specifically from the rhythm in which we type, for how long we hold the keys down on our smartphone or computer keyboard, and the succession of keys used overall in the typing process.

In this absolute regard, keystroke recognition can actually be considered to be the oldest biometric technology around, even much more so than hand geometry recognition or fingerprint recognition. This is primarily so because the interest in unique typing patterns dates all the way back to the nineteenth century, when the Morse code first came out.

By World War II, the U.S. military intelligence department could actually identify enemy Morse code operators by their unique typing patterns.

Although the Morse Code is only, technically speaking, a series of dots and dashes, some distinctiveness could still be established.

The first keystroke recognition device came out in 1979, and by 1980, the National Science Foundation scientifically validated it the technology of keystroke recognition. By 2000, it was finally accepted as a commercial biometric technology that could be used in either the public or private sector.

Keystroke Recognition: How It Works

To start the enrollment process, an individual is required to type a specific word or group of words (text or phrases). In most cases, the individual's username and password are used. It is very important that the same words or phrases are used during both the enrollment and the verification processes. If not, the behavioral typing characteristics will be significantly different, and, as a result, a mismatch will arise between the enrollment and verification templates.

To create the enrollment template, the individual must type his or her username and password (or text/phrase) about 15 times. Ideally, the enrollment process covers a period of time, rather than taking place all at once. This way, the capture of behavioral characteristics will be more consistent.

With keystroke recognition, the individual being enrolled should type without making any corrections (for example, using the backspace or delete key to correct any mistakes). If the individual does make corrections, the keystroke recognition system will prompt the individual to start again from scratch. The distinctive, behavioral characteristics measured by keystroke recognition include

1. The cumulative typing speed
2. The time that elapses between consecutive keystrokes
3. The time that each key is held down (also known as the dwell time)
4. The frequency with which other keys, such as the number pad or function keys, are used
5. The key release and timing in the sequence used to type a capital letter (whether the shift or letter key is released first)
6. The length of time it takes an individual to move from one key to another (also known as the flight time)
7. Any error rates, such as using the backspace key

These behavioral characteristics are subsequently used to create statistical profiles, which essentially serve as enrollment and verification templates. The templates also store the actual username and password. The statistical profiles can be either *global* or *local*. Whereas a *global* profile combines all behavioral characteristics, a local profile measures the behavioral characteristics for each keystroke.

The statistical correlation between the enrollment and verification templates can subsequently be modified, depending on the desired security level. An application that requires a lower level of security will permit some differences in typing behavior. However, an application that requires a higher level of security will not permit any behavioral differences.

It is important at this point to make a distinction between static and dynamic keystroke verification. In the case of the former, verification takes place only at certain times—when the individual logs in to his or her computer, for example. With the latter, the individual's keystroke and typing patterns are recorded for the duration of a given session.

Keystroke Recognition: Advantages and Disadvantages

Keystroke recognition has several strengths and weaknesses. Arguably, its biggest strength is that it does not require any additional, specialized hardware. As previously indicated, keystroke recognition is purely software-based, allowing the system to be set up very quickly.

Second, keystroke recognition can be easily integrated with other, existing authentication processes. The adoption of other biometric technologies requires the implementation of a new process within an existing process. This calls for individuals who are properly trained in the use of contemporary biometric devices, which can greatly increase costs.

Third, everybody is familiar with typing their username and password. As a result, there is very little training required for an individual to use a keystroke recognition system properly. Fourth, the templates that are generated by the system are specific only to the username and password used. Should this username and/or password be tampered with, the individual only needs to select a new username and password to create a new set of enrollment and verification templates.

The weaknesses of a keystroke recognition system are the same as those suffered by other systems that rely on a username/password combination. For example, passwords can be forgotten or compromised while users will have to remember multiple passwords to gain access to, for example, a corporate network. It should be noted that keystroke recognition still requires users to remember multiple passwords (the administrative costs of having to reset passwords will also continue to be incurred).

As such, it only enhances the security of an existing username/password-based system. Second, keystroke recognition is not yet a proven technology. As a result, it has not been widely tested. Finally, keystroke recognition is not necessarily a convenient system to use.

Like signature recognition, keystroke recognition is not widely implemented as some of the other biometric technologies, such as fingerprint recognition and iris recognition. It too can be evaluated against the seven criteria:

1. *Universality*: This is a key strength of keystroke recognition; even people who are *one-finger* typists or not even familiar with typing at all can still be accommodated.
2. *Uniqueness*: At the present time, keystroke recognition only possesses enough unique features to be used for verification applications and not for identity applications.
3. *Permanence*: In this regard, this is one of keystroke recognition's biggest weaknesses, as the typing pattern of an individual can change due to injuries, disease, increased typing proficiency, fatigue, lack of attention, or even using a different keyboard, causing an individual to have a different typing pattern and rhythm.
4. *Collectability*: It can take many typing samples until enough unique features can be extracted.
5. *Performance*: When the proper security threshold setting is established by the systems administrator, keystroke recognition can produce an FRR up to 3% and an FAR up to 0.01%. As mentioned, keystroke recognition does not require any additional hardware, enrollment and verification can happen remotely, it can even be used to further security harden passwords, and the template size is quite small.
6. *Acceptability*: There are no privacy rights issues with keystroke recognition, and there are no negative correlations associated with typing.
7. *Resistance to circumvention*: Any typing data that are not encrypted can be used maliciously by a third party and even be used to spoof the keystroke recognition system. Also, key loggers can be established onto the computer itself to record the various keystroke patterns and rhythms.

Currently, keystroke recognition is included in very few commercial applications. Compared with other biometric technologies, it probably ranks last in terms of use. Having said that, behavioral biometrics are not particularly popular on the whole. Although it is possible that keystroke recognition will establish itself as a dominant technology, there are, at this stage, not enough vendors to propel the technology forward (unlike, for example, fingerprint recognition).

Compared with other physical biometrics, keystroke recognition is easier and cheaper to implement. However, it is unlikely to be used for applications such as physical access control, document verification, passport verification, etc. Instead, it will be used for computer security (where fingerprint and iris recognition solutions are already used as a substitute for usernames and passwords).

Keystroke recognition is also well suited to e-commerce applications. Here, a user would be able to access an Internet banking or e-commerce site by typing in the same text or phrase several times (rather than having to remember different usernames and passwords). Moreover, the same text or phrase can be used to log in to multiple e-commerce sites. Keystroke recognition could additionally be the security tool of choice for multimodal security applications, where it can be used to provide third-, fourth-, or even fifth-tier security.

Although small- to medium-sized enterprises (SMEs) will probably not adopt keystroke recognition, it is well suited to large businesses and organizations, including major banks and financial institutions. It is also quite conceivable that keystroke recognition will be adopted by governments around the world.

BIOMETRIC TECHNOLOGIES OF THE FUTURE

In summary, thus far, we have reviewed all of the physical and behavioral biometrics, with some of the biometric technologies being seen in more commercial usage than the others. However, there is always continuous research and development always going on in biometrics. The following are the biometric technologies that show the greatest promise thus far in terms of development and potential use:

1. DNA recognition
2. Earlobe recognition
3. Gait recognition

The use of DNA as a means of being utilized as a unique piece of evidence did not start until 1984, pioneered by a scientist named Alec Jeffreys. It did not make its mark in the United States until at least the late 1980s, when the FBI invested much resources in exploring the use of DNA as a means of criminal evidence that could be presented in a court of law.

As a result of this massive effort, one of the world's largest databases was born. This is known today as the Combined DNA System, or CODIS. This database is now used by law enforcement agencies at all levels throughout the United States as well as at the international level. In 2012, this database held the DNA records of more than 1.2 million individuals, primarily those of criminals, suspects, and illegal immigrants.

At the present time, DNA is only used for forensic purposes, often collected as a means of latent evidence. Currently, it can only be used for identification purposes and not for verification applications because of the lengthy processing of the collected raw data, which takes about 4 to 5 hours to complete.

However, due to the uniqueness and the richness of the data that DNA possesses, there is thus a strong interest in its use as a potential biometric technology. The use of DNA makes strict use of the patterns that are found in deoxyribonucleic acid and are located in the cell nucleus of all living beings, no matter how large or small.

DNA RECOGNITION

DNA Recognition: How It Works

DNA profiling uses common, repetitive patterns of the four base pairs, which are adenine (A), cytosine (C), guanine (G), and thymine (T). It should be noted that DNA profiling occurs in the uncoded portion of the DNA sequence, which is simply known as the uncoded DNA, for which there is no identified function. A base-pair pattern can look like AGCCTTCAGTA.

This sequence is known as a short tandem repeat, or an STR. A unique STR exists when two or more base pairs appear in a certain and repetitive pattern and also when this exact pattern is repeated closely in adjacent spots. It should be noted that the STR is typically anywhere from two to four nucleotide base letters, and even larger patterns, known as variable number tandem repeats, or VNTR, can contain up to 80 letters. This thus provides for a much higher degree of accuracy and probability. However, the analysis can take up to days to complete, not just a matter of a few short hours.

There are two kinds of DNA that are currently present in all living beings:

1. *Nuclear DNA*. This is the contribution of genetic components from both the mother and the father and is most commonly found in the blood, saliva, semen, and even the bones.
2. *Mitochondrial DNA*. This is the contribution of genetic components from the mother only and can also be found in the hair and the teeth. This is the type of DNA that is used to identify missing individuals.

The raw DNA sample can be collected from blood, semen stains, saliva, hair, or even under voluntary conditions. With the last approach, a buccal swab can be very easily inserted into the mouth to collect a sample of DNA. After the DNA sample is collected from the specific source, the DNA is then isolated and then further divided into much shorter segments (also known as restriction enzymes), which terminate the DNA strands at various points and differentiated in size using an electric current.

The differentiated sizes are then placed onto a nitrocellulose filter, which possesses different fluorescent dyes, which then further attach themselves to the differing repeating patterns. This is then x-rayed and the resultant image is the DNA fingerprint. An image of a DNA strand can be seen in the following figure.

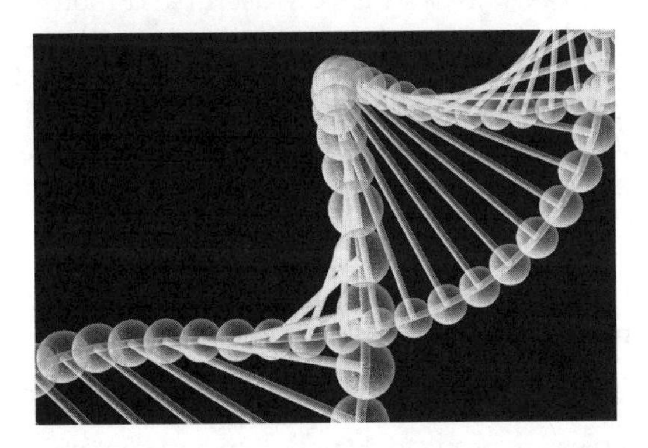

Subsequently, this DNA fingerprint is then translated into a DNA profile that displays the number of STRs at specific locations. The DNA markers used to construct this DNA profile are located on chromosome numbers 1, 2, 4, 5, 10, and 17. These specific locations are referred to as

DNA loci. After this profile is constructed, the findings occur in one of the following three categories:

1. *Exclusion*. This result possesses the highest level of certainty (meaning, the raw data are different from the source from which they were collected).
2. *Nonexclusion*. The raw data cannot be excluded if coming from the same source (in other words, the raw DNA sample and the source from which it was collected are the same).
3. *No result*. No conclusions can be drawn.

DNA Recognition: Advantages and Disadvantages

Because DNA recognition is still under heavy research and development, it can only be matched up against five of the seven criteria, which are as follows:

1. *Uniqueness*: Among all of the other biometrics, DNA possesses the most unique and richest information and data (even much more so than the retina), and if it proves to be a viable biometric technology, it will far surpass the levels of the other biometrics that currently exist in the marketplace.
2. *Collectability*: The right quality and quantity of DNA must be present, and the material must not be contaminated.
3. *Performance*: DNA analysis is very useful and robust for forensic types of applications, but for verification purposes, it is still not useful because analysis takes 4 to 5 h, which must be brought down to a matter of mere seconds.
4. *Acceptability*: DNA is very much prone to privacy rights issues, and while the DNA profile itself does possess genetic information about an individual, the raw DNA data do have this.
5. *Resistance to circumvention*: Although DNA is rich in terms of data and information, it can often be very easily substituted for another person's DNA.

Other disadvantages of DNA recognition include the following:

1. DNA samples are prone to degradation and contaminants.
2. Implementing the security protocols for a DNA-based biometric system could prove exceptionally complex. This applies to database protection and overall system confidentiality in equal measure.

GAIT RECOGNITION

Gait recognition is the next potential biometric that is being looked at very closely as a viable technology into the future. Gait recognition, simply put, is the potential verification and/or identification of an individual based upon his or her unique stride or the unique way in which one walks in normal, everyday life. It can be considered both a behavioral- and a physical-based potential biometric and thus is considered to be part of the field known as kinematics, which is the study of motion.

With our stride, the various aspects of the body closely position themselves at unique lengths, angles, and speeds. At the present time, there is strong interest in using gait recognition for identification purposes because the raw data that are needed to create the biometric templates can be collected in a very covert fashion. As a result, gait recognition can be used very well with CCTV technology and facial recognition.

The Process Behind Gait Recognition

There are currently two methods that are being investigated for capturing and examining the unique strides in individuals:

1. *Static shapes*. This methodology actually takes a silhouette pattern of a walking person from a series of video frames taken by a CCTV camera. This image of the walking individual is defined in a boxed region of the video frame, and any extraneous features from the external environment are removed. The silhouette image is then created, and from there, a further series of images are created and then digitally mapped to yield what is known as the gait signature.
2. *Dynamic shapes*. This type of methodology uses an accelerometer, which is placed on the individual's leg, and the stride (walking) motion is then collected and analyzed from three different directions.

There are numerous methods that are currently being investigated to capture the raw data:

1. *Machine vision*. With this, the individual video frames that display the individual walking are isolated from one another frame by frame.
2. *Doppler radar*. With this system, a sound wave is bounced off of the leg of the individual, and the return sound wave is then

measured. This is done by ascertaining the velocity of the signal that is modulated by the moving parts of the leg.
3. *Thermography.* This technique is embedded into a computer vision system to cast out the moving leg in difficult and extraneous lighting situations.

After the raw images are captured and collected, the next step in the biometric process is unique feature extraction. Currently, there are two methodologies that are also being examined to accomplish this task:

1. *Model-based analysis.* With this technique, the height of the individual, the distance between the head and the pelvis, the total distance between the pelvis and the feet, as well as the total distance between the moving feet are measured and calculated. Also, the relationships of the distances and the angles of the walking individual can be converted into various vectors as well.
2. *Holistic analysis.* With this technique, the silhouette of the walking individual are also calculated and measured.

Finally, after unique feature extraction, the next step is the creation of the biometric templates. In this regard, gait recognition is different from the other biometric technologies in that actual sequences of movements are compared against one another. There are also various template creation techniques that are under research and development:

1. *Direct matching.* With this approach, a gait sequence is identified by comparing the raw data that are captured with the reference data, and from there, the various distances that yield the least minimal distances are subsequently calculated.
2. *Dynamic time warping.* With this technique, the closeness between two sequences of gait movements are measured. The variables here examined and analyzed include the time, speed, and various types of walking patterns.
3. *HMMs.* This is a statistical-based methodology in which the probability of the similarity of the shapes that appear in the walking stride of an individual is examined and analyzed in succession.

Gait Recognition: Advantages and Disadvantages

Gait recognition, although still under research and development, can be compared and evaluated against the same seven criteria the other biometric technologies have been rated:

1. *Universality*: The collection of raw data requires that an individual possess a walking stride that is basically unimpeded and be without the use of such walking aids such as crutches, walkers, and even wheelchairs.
2. *Uniqueness*: Since gait recognition is not yet a viable biometric technology, its true, absolute uniqueness still cannot be ascertained yet when compared with the other biometric technologies, and as a result, at the present time, gait recognition is predicted to work very well in a potential multimodal biometric security solution.
3. *Permanence*: A person's walking stride can be affected by such variables as older age, physical injuries, or even disease. The walking stride can also be affected by such variables as the type of surface the individual is walking upon, any additional or extraneous weight, which is being carried by the individual, and even the type of footwear the individual uses while he or she is walking.
4. *Collectability*: A potential advantage of gait recognition is that raw data that have a low resolution as captured by the CCTV system can still be used to create robust biometric templates.
5. *Performance*: Some of the studies conducted thus far have claimed that gait recognition possesses an ERR of 5% (and perhaps even less).
6. *Acceptability*: It appears that gait recognition can be considered to be potentially a noninvasive biometric, but the issues of privacy rights and civil liberty violations could emerge as a person's unique walking stride can be collected and analyzed covertly.
7. *Resistance to circumvention*: At this point in research and development, it is deemed that it could be very difficult to mimic another individual's walking stride.

Finally, it is forecasted that gait recognition can be used as a great surveillance tool (especially when combined with CCTV technology and facial recognition) in large venue settings, such as sporting events, concerts, transportation hubs, educational institutions, etc. Also, gait recognition can be used to track suspicious behaviors with regard to motion of an individual, for example, if a particular person appears in the CCTV field of view an unusual number of times in a short time frame.

EARLOBE RECOGNITION

Earlobe recognition is the last potential biometric technology to be examined in this chapter. The examination of the unique features of earlobe recognition dates all the way back to the nineteenth century, and an ear classification system was first proposed way back in 1964 by the scientist Alfred Iannerreli. In the mid-2000s, the European Commission further studied the use of the ear as a potential biometric.

Earlobe recognition is a potential physical biometric, in which the unique features are extracted and examined. The ear is fully shaped by the time an infant turns 4 months of age. As a potential biometric, it is the structure of the outer ear that is of prime interest. An example of the outer ear can be seen in the following figure.

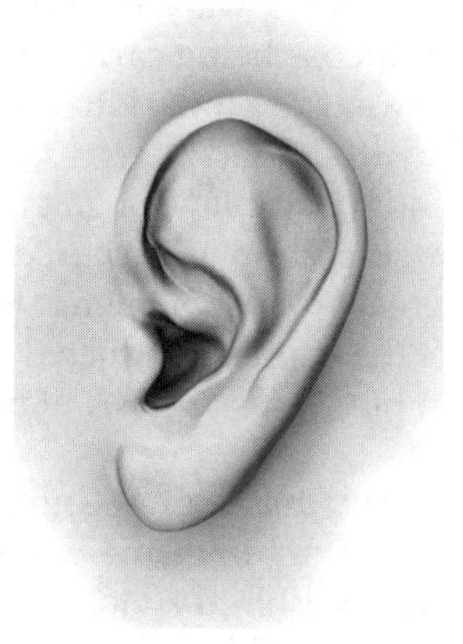

Earlobe Recognition: How It Works

Current research is focused upon using a technique known as *eigenears*, which is based upon the different structures of the outer ear. Eigenears is very similar to the technique of *eigenfaces* used in facial recognition. Thus far, there are positive results in the scientific community that the outer ear is unique in its overall shape and in terms of depth, angular structure, and the overall formation of the earlobe structure.

From this, scientists are currently examining the distances between the regions of the ear, in 45° and 3-D angular models. The raw data of the ear can be captured by taking an actual picture of the outer ear, preferably when the ear is placed against a glass platen to capture the actual, raw images. These raw images are then used for unique feature extraction and biometric template creation. Either a 3-D or a thermal picture can be taken of the outer ear, with preference being given to the latter.

At the present time, there are two methodologies that are currently being examined for potential earlobe recognition:

1. *PCA*. With this technique, the raw image of the ear is cropped; it is then scaled to a regular size, thus capturing the two key-points of the ear, specifically the regions known as the triangular fossa and the antitragus, from which the unique features can be extracted.
2. *3-D analysis*. With this technique, the outer image of the ear is captured, thus ascertaining both the depth and the color type of the ear. Both the helix and the antihelix are examined, from which the unique features can be captured.

Earlobe Recognition: Advantages and Disadvantages

The potential of earlobe recognition can also be examined from the same seven criteria against which all of the other biometric technologies in this chapter have been analyzed:

1. *Universality*: For the most part, unless an individual has a physical injury that caused him or her to have an absence of the ear, most people have an ear that can be examined for unique features.
2. *Uniqueness*: Although it is still not yet proven scientifically, in the two tests involving 10,000 earlobes and identical twins, preliminary results have shown that all individuals were found to each possess a unique ear structure.
3. *Permanence*: The shape of the ear more or less stays the same over the lifetime of an individual.
4. *Collectability*: Certain variables from the external environment can certainly impact the capture of the raw images of the ear, and these variables include different types of lighting situations, jewelry such as the wearing of earrings, and wireless device accessories that can be placed in the outer ear.
5. *Performance*: It is still too early to determine how well earlobe recognition will actually be adopted and used in the commercial market.

6. *Acceptability*: Earlobe recognition is deemed to be a potentially nonintrusive type of potential biometric technology; thus, there should be no issues with regard to privacy rights.
7. *Resistance to circumvention*: At this point, it is difficult to determine this, but it is possible to construct a false earlobe.

Although there are no commercial applications yet of earlobe recognition, in the future, it is heavily anticipated that it will work well with facial recognition systems in multimodal biometric solution applications.

Finally, of the three potential biometric technologies covered in this chapter, gait recognition appears to offer the most potential in terms of practical deployment. Its potential is very strong for multimodal security applications. Think, for example, of a dual-level verification system whereby a subject's identity is initially established based on, say, a fingerprint before being verified using gait recognition. It is also useful for one too many verification requirements at, for example, airports.

Under these conditions, gait recognition can offer significant time savings by allowing large groups of people to be identified in a single environment. Compared with gait recognition, earlobe recognition has much further to go, also in terms of research and development. This also applies to DNA recognition, even though it has the potential to be the best biometric of all (on account of its uniqueness).

REVIEW OF CHAPTER 2

In review, Chapter 2 examined the biometric technologies that are in use today and what is being planned for the future. More specifically, the physical biometric technologies were examined (which include hand, finger, retinal, iris, vein pattern, palm print, and facial recognition) and the behavioral biometrics were also reviewed (which include keystroke and signature recognition). Also, the biometric technologies of the future were also looked at (which include DNA, earlobe, and gait recognition).

Of all of these biometric technologies, it is primarily fingerprint, iris, and VPR that are currently being deployed the most. For the future, it is gait recognition that holds the most promise of being a viable biometric technology. This fact brings up a very important subject that is very important to any CIO as well as any other C-level executive. That is the overall topic of biometric system analysis and design.

For example, before any biometric technology can be considered for financial acquisition and ultimate deployment, the existing security infrastructure of the place of business or organization must be very carefully mapped and examined to fully determine just how well a particular biometric technology can fit into it.

For instance, key questions must be carefully evaluated and answered. Some of these questions include the following:

1. Will the particular biometric technology being considered be deployed as a stand-alone system (in other words, will it be the primary means of defense for the business or organization)? Or, will it be implemented as a security add-on within the existing security infrastructure (such as a multimodal security solution, where there are multiple layers of security)?
2. Will the biometric technology being considered for deployment be wired and networked within the existing security infrastructure? Or, will a different networking environment be needed?

The goal of this chapter is to explore, examine, and provide answers to these particular questions and much more. This chapter is intended to be an introductory project tool for C-level executives, as they contemplate the choice of biometric technology that will be the most optimal and work best for their place of business or organization.

3

For the C-Level Executive

A Biometrics Project Management Guide

Specifically, this chapter is divided into the following topics:

1. Biometric technology system architecture
2. Biometric system analysis and design processes (the biometrics project management section)
3. Biometric networking topologies

This chapter is not intended to be a substitute for professional project management advice. Rather, the goal is to give the C-level executive the starting point he or she needs in which to determine the serious questions that need to get asked (and answered) to get the maximum use of his or her security budget.

BIOMETRIC TECHNOLOGY SYSTEM ARCHITECTURE

Generally speaking, biometric devices, no matter how large or small they might be, are actually quite complex, even though that may also look very simple in nature on the outside in terms of design and ergonomics. However, the exact technology components will vary from vendor to vendor, but all biometric technologies (even whether it is physical- or behavioral-based) consist of the following subcomponents:

1. Sensing and data acquisition
2. Signal and image processing

3. Data storage
4. Template matching
5. Threshold decision making
6. Administration decision making
7. Data (biometric template) transmission

SENSING AND DATA ACQUISITION

Each of these subcomponents will be reviewed in detail in this chapter, first starting with the sensing and data subcomponent. In its simplest terms, this subcomponent involves the capture and the extraction of raw biometric data (in particular the image), and from that point onward, converting those raw data (the image) into a form that can be used for further processing from within the particular biometric technology.

This subcomponent can collect two types of raw data, and this happens at first in the enrollment stage and then at the secondary verification and/or identification stage. These two types of raw data that can be collected are defined as follows:

1. *Probe data*: These are the first series of raw data to be collected, and they are used to create the enrollment template.
2. *Live sample data*: These are the second series of raw data to be captured, and these are used to create the subsequent verification templates (which in turn are then used to compare against the enrollment template for various types of verification and/or identification applications).

Since this subcomponent primarily deals with the collection of biometric raw data, the way, manner, or technique in which either the physical or the behavioral biometric is presented to the sensing device thus becomes of prime importance. Specifically, in this regard, the term *presentation* can be defined as "the submission of a single biometric sample on the part of the user." The factors that affect presentation can be specifically defined as "the broad category of variables affecting the way in which the users' inherent biometric characteristics are displayed to the sensor" (*Certified Biometrics Learning System, Module 2, Biometrics System Design and Performance*, © 2010 IEEE, pp. 3–4).

Therefore, the standardization process of presentation also becomes of prime importance and involves ensuring the following to establish a consistent, raw sample acquisition process:

1. The ease of use in ergonomic sensor design.
2. Effective user and operator/administrator training (extreme familiarity with the biometric system is an absolute must here).
3. Quality feedback mechanisms put into place for both the end user and administrator so that the presentation technique can be constantly improved upon.

Multimodal Biometric Systems

In today's biometric deployments, rarely do you find just one device being used. Rather, you see multiple devices being used, thus requiring multiple sensing devices. In this regard, the following sensor schemes can be established:

1. One sensor of multiple devices of the same biometric modality
2. Multiple sensors of multiple devices of differing biometric modalities

Such designs and layouts are known as *multimodal biometric solutions.* In this regard, there are two different types of biometric multimodal systems:

1. Synchronous systems
2. Asynchronous systems

Synchronous means that there are two or possibly more biometric systems being used at the same time, for the same authorization process of the subject in question. In this scenario, imagine a fingerprint scanner and an iris scanner being used simultaneously. *Asynchronous* means that there are two (again possibly even more) biometric systems that are being used, but the key difference here is that they are used in tandem with each other, or in a sequential fashion. Again imagine the fingerprint scanner and the iris scanner. The fingerprint scanner would be used first, and immediately after, the iris would be scanned to confirm the identity of the subject in question.

There are two key points to remember when using biometric multimodal systems:

1. Using multiple systems increases the accuracy, although it does not increase the speed of the authorization process.
2. If one device does not work, the other systems will come into play as a fail-safe, thus still assuring a strong level of security.

With regard to physical access entry, it is typically businesses and organizations that use multimodal biometric solutions. Very often, there will be a strong presence of security at the main points of access and entry of a building, and within the office infrastructure itself, other rooms and private areas will be protected also. At the main points of entry, a hand geometry scanner is used to verify and confirm the identity of individuals.

Along with this scanner, there could be other levels of security used in conjunction, such as a swipe card or smart card system. When the individual gains entry through the main point of access, the internal parts of the office remain protected, such as the client file room and the server room. Very often, fingerprint recognition is used in this extra layer of security. This then formulates the multimodal biometric solution in this example, and as you can see, it is an asynchronous approach. Typically, this combination of security is a very powerful one. One of the most powerful forms of a multimodal security solution is combining CCTV technology with facial recognition, as shown in the following figure.

Also, other types of biometric devices can be used in lieu of fingerprint recognition, such as facial and iris recognition. Nowadays, VPR is starting to be used extensively as a component of a multimodal biometric package.

Single Sign-On Solutions

In terms of logical access entry, passwords have long been the traditional means of security when logging on to a computer or a corporate network.

However, as has become clear with the proliferation of the identity theft threats, passwords can be cracked and hacked very easily.

Thus, organizations and businesses are constantly faced with enforcing stronger password policies, frustrating the end user even more. However, even here, biometrics is emerging as a much favored security solution over the password. Nowadays, you can log into your network or computer with the single swipe of a finger or even a picture of your iris.

Today, new laptops often contain a small fingerprint sensor on the keyboard. Although durable, significant damage can occur to the sensor if the laptop is dropped or exposed to harsh environmental conditions. To counter this, a secondary biometric system can be used, such as iris recognition. For example, when logging on to their laptop, an individual's identity could first be confirmed with a quick swipe of the finger, and then for extra protection, his or her iris could be scanned and verified via a built-in camera, before full access is granted. An example of this is portrayed in the following figure.

As is the case with physical access entry applications, here also VPR is emerging as the top choice in multimodal biometric solutions. In the world of logical access, the cardinal rule of having two or more means of identity verification is just as important, if not more.

Implementing a Multimodal Biometric System

In the past, implementing a multimodal biometric system was an expensive endeavor, as a biometric system would have to be implemented on top of, or with, an existing legacy security system. This would very often mean a major overhaul in the organizational infrastructure. However, the past decade has seen an evolution in biometrics technology as well as in multimodal systems.

One unit or device can now house different technologies. Nowadays, multimodal biometric solutions are not just available as hardware but can be obtained as a software development kit (SDK). This approach has become known as a *hybrid* biometric system, where fingerprint, iris, and VPR are primarily used.

Before implementing either a hardware or a software multimodal biometric system, there are some key questions that need to be taken into consideration:

1. As a business owner, are you trying to protect the physical or the logical aspects of your business?
2. Which biometric system will work best for your particular multimodal system?
3. Will the new multimodal biometric system be layered with the existing security infrastructure or will it be used as an addition?
4. What are the costs, and what is your budget?
5. What kind of software applications, if any, will need to be developed to support a new multimodal biometric system?

After evaluating some of these major factors, it will soon become apparent that a multimodal biometric solution offers a number of strategic benefits, which include the following:

1. A fortified level of security. Rather than having just one layer of security, you will now have multiple layers. Not only will there be a fail-safe mechanism in place, but also should a perpetrator break through one line of defense, they will most likely be caught deeper in. In other words, an SME can retain a very high security threshold with this scheme.
2. There are many metrics that are included in evaluating the effectiveness of a biometric system, for example, the false accept rate (FAR), the FRR, and the ERR. Using a multimodal biometric system, the reliability of the verification process is greatly increased.
3. Some systems experience difficulties enrolling and verifying people of different ethnic origins, but a multimodal biometric system can capture the unique characteristics of a much larger and varied population. A read rate of almost 100% can be guaranteed.
4. In biometrics, there is always the theoretical issue that a system can be spoofed. By having extra layers of biometrics, this issue has become, over time, much less prevalent.

5. As mentioned previously, a newer version of the multimodal system is known as the hybrid approach. This permits for more than one biometric technology to be used within one component. An example of this is an SDK, which allows for the different recognition technologies to all be used together, independent of the actual hardware being utilized.

Challenges with Multimodal Biometric Systems

One of the biggest challenges to the sensing subcomponent, especially in multimodal biometric systems, is that of the physiological or behavioral trait that is presented to the biometric sensor, and the raw image that is subsequently extracted changes each and every time. In other words, there is no such thing as two 100% identical biometric templates that are created in the end.

There are always changes and variability that occur based upon a number of key factors:

1. The actual physiological or behavioral trait has been changed over a period of time (such as due to injuries, aging, etc.).
2. The way in which the physiological or behavioral trait is presented to the biometric sensor changes over time on the part of the end user.
3. The biometric sensor degrades over a period of time as well, due to everyday wear and tear or even to technological breakdown.

Thus, as one can see, the way and manner in which the physiological and/or behavioral trait is presented to the biometric sensor on the part of the end user has a direct impact upon biometric system performance and the quality of the enrollment and verification templates that are subsequently created.

The following variables are deemed the most important in the sensing/data acquisition subsystem phase:

1. *Collectability*: If the physiological or behavioral trait cannot be measured by the biometric system, no matter how good the particular trait may be, it will be deemed to be useless.
2. *Performance*: The quality of both the biometric sensor and the raw image must reach or even exceed the required performance levels.

3. *Acceptability*: The method of sensing must prove to be very easy to use and hazard-free by the end user.
4. *Circumvention*: The biometric sensor must be 100% *spoof-proof.*

A function that is very important in presenting a physiological or behavioral-based biometric trait to the biometric sensor is that of the attempts made by the end user. Specifically, as it relates to biometrics, an attempt can be described as "the submission of a single set of biometric samples to a system for identification or verification. Some biometric systems permit more than one attempt to identify or verify an individual" (*Certified Biometrics Learning System, Module 2, Biometrics System Design and Performance*, © 2010 IEEE, pp. 3–7).

Note that there are two specific types of situations that can lead to a repeated number of biometric trait attempts by the end user:

1. The biometric trait that is presented to the sensor is rejected because it is deemed to be totally unreadable.
2. The biometric trait that is presented to the sensor is ultimately accepted and processed successfully into a verification template, but a nonmatch is still declared by the biometric system because there is no current enrollment template that exists for the user in the system's database.

The biometric system can keep taking samples of the physiological or behavioral trait literally forever. If this occurs, the end user will probably just quit the entire process of verification and/or identification all together. The end result will be that the end user will be noncooperative for future verification and/or identification attempts that will be required of him or her to gain access to the resources he or she is requesting.

Therefore, it is very important for the system administrator to set the appropriate quality threshold of the biometric system so that a balance is struck in the number of attempts that are required by the end user. In other words, it is a fine line that needs to be maintained between end user perception and convenience versus the security of the business or organization.

However, this balance comes with a sharp trade-off:

1. Requiring more attempts on the part of the end user will increase both the FAR and the FER.
2. Requiring a fewer number of attempts on the part of the end user will greatly increase the false match rate (FMR) and the false nonmatch rate (FMNR).

Finally, there are three other factors that affect the sensing and data acquisition subcomponent:

1. *Illumination*: This is the dedicated light source that exists from within the biometric device. This light source flashes onto the physiological trait so that the raw image can be captured, but the amount and intensity of light used vary greatly from technology to technology (a lot of this is due to the compensation for the extreme changes in lighting in the external environment, in order to capture a consistent, quality raw image).
2. *Contrast*: This refers to the differentiation that exists between the lightest and the darkest portions of the biometric raw image, which is ultimately captured. It can also refer to the total number of dark pixels versus the total number of light pixels in the raw image.
3. *Background*: All biometric systems keep the background image to a solid, light color.

SIGNAL AND IMAGE PROCESSING

The next subcomponent to be examined after sensing and data acquisition is the signal and image processing subcomponent. This is the part of the biometric system that consists of highly specialized mathematical algorithms that detects, enhances, and extracts the unique features from the raw image that was captured in the previous step just described.

These very sophisticated mathematical algorithms take these unique features and convert them to the appropriate mathematical file type. This in turn becomes the enrollment and the verification templates. The signal and image processing subcomponent consists of the following steps:

1. Preprocessing of the biometric raw image
2. Various types of quality control checks
3. Biometric raw image enhancement
4. Unique feature extraction from the raw image
5. Postprocessing of the unique features
6. Unique feature compression and decompression

Preprocessing of the Biometric Raw Image

The first portion of the signal and image processing step is the actual preprocessing phase. In this phase, the biometric raw image is automatically fit into a standardized orientation, the center point is identified, and any extraneous data are segmented out and fit to be contoured into a certain set of established permutations.

Within the preprocessing phase, there are numerous subphases that occur:

1. *Detection*: This is the determination of the type of biometric signal that exists in the data stream, which is captured from the biometric raw image. For all of the existing biometric modalities (both physical- and behavioral-based), the detection of sort of a signal will occur because the physiological/behavioral trait is presented consistently and voluntarily to the sensor.
2. *Alignment*: This involves determining the proper orientation and centering of the biometric raw image. With the former, this includes taking into account any types of differences and variances in the specific rotation of the biometric raw image. This step is very important, as it can increase the accuracy of the unique features that are captured as well as the quality of the subsequent enrollment and verification templates that are created. With the latter, the center of the biometric raw image is ultimately determined, and based on this, the image is then repositioned.
3. *Segmentation*: This part of the preprocessing phase can be specifically defined as "the process of parsing the biometric signal of interest from the acquired data system" (*Certified Biometrics Learning System, Module 2, Biometrics System Design and Performance,* © 2010 IEEE, pp. 3–11).

Before the segmentation subphase can occur, the detection subphase must be accomplished first successfully. After all, to actually segment the biometric raw image, there must be proof of actual data in it, which is what the detection subphase exactly accomplishes.

From the biometric raw image, once the needed data are located, the extraneous data are then cropped into the segmentation subphase. This subphase can also be applied to video sequences as well, especially when facial recognition is being used in conjunction with CCTV technology.

For example, only the relevant scenes where the face distinctly appears can be included, with the other irrelevant scenes being cropped out appropriately. The next subphase after segmentation is what is known as the normalization subphase. This step involves adjusting or scaling of the relevant biometric data that are discovered in the last subphase. This is then contained within a range of predetermined values that are deemed to be acceptable.

Normalization can literally increase or decrease the levels of a signal that has been detected previously and also enhance that particular signal so it can be processed more efficiently by the biometric system in subsequent steps. There are two different types of normalization:

1. *Energy level normalization*: This process subtracts the effects of differing levels of luminescence by increasing or decreasing the contrast among the features. Also, this process is usually applied to regions with the detected signal, and grayscale normalization decreases the differences found in the gray levels of the biometric raw image.
2. *Scale/shape normalization*: In this process, the signals are exactly modified so that easy comparisons can be made. Also, biometric raw images that have complex shapes or curvatures can also be adjusted with normalization.

Quality Control Checks

The next subphase in the signal/image processing is quality control, and it consists of the following components:

1. Assessing the level of quality in a feedback loop
2. Using quality control as a predictive mechanism for measuring how well biometric templates will compare with another (in particular, the enrollment template versus the verification template)
3. Assessing the level of quality control as an input mechanism from within multimodal biometric solutions

In the first component (quality control in a feedback loop), it is very important that the biometric system detect any flaws in the signal that is captured in a very quick and efficient manner. In this particular instance, if a poor signal is determined, the feedback loop mechanism in the biometric system is automatically triggered and the end user is asked by the system to present their particular biometric trait again.

This particular mechanism is especially useful for the noncooperative type of an end user, whose biometric trait has to be presented and repeated numerous times.

With the second component (quality control as a predictive mechanism), this is very often used to gauge how well the two templates (enrollment and verification) will compare against one another. In these types of scenarios, very often, quality scores are calculated and utilized. The quality score serves four distinct purposes:

1. If the signal can provide enough data so that the two templates can be matched against one another.
2. The quality control score can also be utilized as an indicative tool to measure and ascertain just how much signal and image processing is really required by the biometric system.
3. The quality control score can be used to formulate a specific type of matching strategy (for example, if both sets of irises will be scanned or just one iris will be scanned or if two fingers will be scanned versus all ten fingers, etc.).
4. The quality control score can serve as a measurement technique as well to determine if the needed or required amount of unique features have actually been extracted to create quality enrollment and verification templates.

In the third component (quality control as an input mechanism), the quality control score can be used to *meld* and bring together biometric information and data from the various biometric modalities that exist in a multimodal biometric system.

In such solutions, quality control scores can also be utilized to create different weighted values in the signal and imaging subphase of the different biometric modalities, which are involved in the overall security solution.

Image Enhancement

The next subphase after the quality control subphase in signal and image processing is the enhancement subphase. This involves taking the signal and enhancing and/or further processing it for further use later by the mathematical algorithms that create the enrollment and the verification templates.

The enhancement subphase involves the following components:

1. The replacement of the unreadable features of the biometric raw image with blank spots, so that only clean and distinct unique features will be extracted.
2. Taking the very complex parts of the biometric raw image and simplifying them further for easier processing by the biometric system.
3. The fixing and the repairing of any damaged or missing features from the biometric raw image. In other words, this is the extrapolation and calculation of what a feature could very possibly look

like, based upon the careful examination and analysis of adjacent features (this is a must—the degraded feature(s) must be next to or near a reliable and distinct feature)

Feature Extraction

After the enhancement subphase in signal and image processing comes the feature extraction subphase. This is also known as representation, and this is the very core of it. Feature extraction occurs at both the enrollment and the verification and/or identification stages.

It consists of the following methodology:

1. Unique feature location and determination
2. Extraction of the unique features
3. Development and creation of the enrollment templates
4. Behavioral model creation
5. Development and creation of the verification templates

With unique feature location, this simply involves the examination of the biometric raw image of the distinct features that sets them apart from the other features. For example, with a fingerprint raw image, this would involve the location of the minutiae within the ridges and valleys, and with an iris raw image, this would involve the location of the furrows, coronas, freckles, etc. that exist from the iris itself.

Extraction simply involves, via the use of the appropriate algorithm, the removal of the unique features that will be utilized. With verification template creation, this simply involves taking the unique features that were removed previously and converting them over to the verification template, which is nothing but a mathematical representation of the unique features. For example, with the minutiae of the fingerprint, it would be represented by a binary mathematical file (which is a series of zeroes and ones).

In terms of behavioral model creation, statistical models are created to represent the unique features. As its name implies, this is used primarily for the behavioral-based biometrics of signature recognition and keystroke recognition, where there are no physiological raw images to be captured; behavioral-based biometrics are much more dynamic in nature-versus physical-based biometrics, where the raw images are much more static in nature.

Finally, with the creation of the enrollment templates, the same processes are used in the creation of the enrollment templates, but this time, the verification template is compared with the enrollment template to determine the closeness of the two templates. Subsequently, the result of this comparison is used to either grant or deny access to the individual in question who is requesting the use of a particular resource.

Specifically, it is the feature extraction algorithms that take all of the unique data that are extracted and convert them over to the appropriate mathematical files, which thus become the enrollment and verification templates. Although the goal here is to maximize the time spent on determining and ascertaining the distinctiveness of a biometric raw image and to render less time on the extraction and matching processes, there is a distinct trade-off to be made.

For example, the less time that is used in analyzing the distinctiveness of the biometric raw image, the more time that is needed for unique feature extraction and matching of the enrollment and verification templates. Finally, a good biometric raw image is one that has a lot of distinctiveness and uniqueness to it and whose physical structure does not change much over the lifetime of an individual.

Postprocessing

The next subphase after feature extraction is postprocessing. This occurs when the unique features are extracted. The postprocessing subphase looks for any groupings of the unique features that may have occurred in the extraction process (any grouped unique features can cause errors and increase the processing time later) or for any poor quality features.

Also, during this subphase, the biometric raw image is either discarded or archived from the biometric system. However, there are issues to be considered. For instance, retaining any type of biometric information and data can raise serious privacy rights issues, and if any of it is discarded by any means, it can cause a serious hindrance for future upgrades in terms of both biometric hardware and software for the future (for instance, if the biometric information and data are indeed discarded, the end users may very well have to re-enroll their templates yet again if the newer biometric system is ever set into place).

Data Compression

The very last subphase in the signal and image processing is known as data compression. This subphase is used primarily in two distinct scenarios:

1. If the biometric template or data take up too much memory in the biometric system
2. If the biometric template or data must absolutely have to travel across a low network bandwidth channel

The basic premise of data compression is to bring down the biometric data template or data into a neat and manageable size while keeping the essential information and data needed for verification and/or identification intact. Data compression is deemed to be successful if the template or data, after they have been decompressed, can function in the exact same manner as they were intended to before they were compressed down.

There are two extremes associated with data compression, and these are known as *lossy* and *lossless*. With the former, this creates an irreversible loss of all of the biometric template information and data, and the latter creates a situation where all of the biometric template information and data can be recovered after they have been compressed down.

It should be noted that most compression methods fall right in the middle of this spectrum. Also, a higher compression ratio will result in a much higher degradation of the biometric template information and data. This is so because some of the biometric information and data are either rounded up or down, and the original information and data are usually not retained. To counter this, neural network technology can be implemented to recreate any lost or damaged biometric template information and data after they have been compressed down.

DATA STORAGE

After the signal and image processing subcomponent comes the data storage subsystem. This subsystem can be viewed as the database of the biometric system and is thus primarily responsible for the storage of all biometric information and data, including both the enrollment and verification templates. Specifically, a biometrics database can be defined as "a collection of one or more computer files. For biometric systems, these files could consist of biometric sensor readings, templates, match results, related end user information, etc." (*Certified Biometrics Learning System, Module 2, Biometrics System Design and Performance,* © 2010 IEEE, pp. 3–24).

A concept associated with a biometrics database is that of a gallery. A gallery can be defined as "The biometric system's database, or set of known individuals, for a specific implementation or evaluation experiment"

(*Certified Biometrics Learning System, Module 2, Biometrics System Design and Performance*, © 2010 IEEE, pp. 3–24). Simply put, a gallery is a database of images, such as that of what a facial recognition system would use (primarily for storing eigenfaces and CCTV camera footage).

Another major function of the biometrics database other than for storage is search and retrieval, which is a direct function of the matching process when the enrollment and verification templates are compared with one another. For example, if identification (1:N) is the primary application scenario of the biometric system, then the entire database must be searched.

In the above type of scenario, a search database is then created from the master biometric database in an effort to expedite the searching process. This secondary database (the search database) is created to protect the master database by searching through its various partitions.

Search and Retrieval Techniques

With a biometrics database, there are two types of search and retrieval that can occur:

1. *Verification*: This is when the database is searched for a claimed identity. This is a very straightforward process because only one matching record has to be found, which in turn leads to less processing time imposed onto the biometric system.
2. *Identification*: This is when the entire database must be searched completely because of the lack of a claimed identity. This can be a very complex process, can be time consuming, and can severely constrain biometric system resources.

Search and retrieval techniques in biometrics-based databases are affected by the following factors. In other words, the following variables can greatly affect the processing of the biometrics database:

1. *Winnowing errors*: Although the advantage of winnowing a database (which involves the filtering, binning, and indexing of records within a database) is designed to optimize the search space in a database (by creating the temporary partitions in it), it can also lead to search and retrieval errors.
2. *Differences in the quality of the biometric data in the biometrics database*: Search and retrieval techniques are also greatly affected by the differences in the quality of data that exist in varying types of multiple databases (for instance, those found in multimodal

biometric systems). To compare data efficiently and successfully in these heterogeneous environments, the biometrics database system must be able to quantify the performance for ensuring good quality comparisons of the differing biometric templates. As a result, image processing algorithms are introduced to help fuse the biometric raw images with the different modalities.

3. *Multiple types of biometric data*: In multimodal biometric systems, multiple biometric templates can be searched for simultaneously. Although this is designed to greatly decrease the search and retrieval time required, it also puts extra burden onto the processing resources of the biometric systems in a multimodal environment.
4. *Biometric template accuracy requirements*: If further granular detail is required of the biometric templates, the biometric system can easily extrapolate such features via the mathematical algorithms it possesses. However, again, this can also increase search and retrieval time within the database and be a further constraint on the processing resources.

Database Search Algorithms

To search a biometrics database effectively and efficiently, various types of search algorithms can be used:

1. *Sequential searching*: This involves examining and analyzing biometric templates for the first through the last records in a chronological or sequential order.
2. *Binary search trees*: With this technique, a directory root tree structure is placed in a specific location within the biometrics database. This thus creates two smaller subtrees. The biometric templates can then be searched for in one subtree or the other subtree.
3. *Hashing*: This method uses a specific key that is assigned to each biometric template in the database to allow for quick searching. This is especially useful in verification (1:1) applications.
4. *Parallel searching*: This is where multiple biometric templates are searched for at the same time in multiple biometrics databases.

Another very important element of the biometrics database is the actual server upon which it will reside and be hosted from. Typically, in small- to medium-sized applications, the actual biometrics database resides within the biometrics database itself. However, in much larger applications, especially that of client–server types, the biometrics database resides on a separate, physical server.

The hardware in which to store the biometrics database are as follows:

1. *Virtual servers*: These are typically stored at the physical premises of the Internet Service Provider (ISP) and are often the most economical choice due to economies of scale.
2. *Blade servers*: This type of server can help with hardware standardization and allow for the biometrics database applications to be scaled upward quite easily and rapidly.

The following are images of virtual and blade servers, respectively.

An equally important component along with the choice of hardware is the software that is used in the design and creation of the biometrics database. A lot will depend upon the requirements of the end user, but most typically, either closed- or open-source database software can be utilized in development.

With the former, the choices range from SQL Server to Oracle database software, and with the latter, the choices range from MySQL to PostGRESql database software. Also, along with the choice of the appropriate hardware and software, the following are equally important factors that need to be taken into consideration before the actual biometrics database is created and built:

1. The requirements of the biometrics systems based upon the system analysis and design
2. The bytes size, the total number, and the types of biometric templates that will be stored into the biometrics database
3. The exact biometric application that will be used (such as either identification or verification)
4. The expected increase in size of the biometrics database (thus, scalability will be very important here)
5. The speed and efficiency of the search and retrieval queries for particular types of biometric information and data
6. Any privacy rights issues that may arise with regard to the storage of the biometric templates in the biometrics database
7. The overall network security of the biometrics database
8. The overall stability of the biometrics database in the long term.

Backup and Recovery of the Database

After the hardware and the software choices have been decided upon to build the biometrics database, the next factor that needs serious consideration is the backup and recovery of the biometrics database, in case of either physical or natural disasters. The considerations in this factor include the following:

1. The detection and repair of any types of errors in the biometrics database
2. The type of schedule implemented for backup of the biometrics database
3. Any offsite storage locations for the biometrics database
4. Any type of redundancy (this primarily includes the choices of both fail safe and fail over options)
5. The backup and recovery plan in case that the biometrics database is either lost or heavily damaged due to physical or natural disasters.

Finally, it should be noted that most biometric systems use a combination of storage mechanisms just described to ensure continued operations in case of either biometrics database failure or network connectivity failure.

Database Configurations

In terms of biometrics database configuration, there are presently four different types of setups that are available:

1. A centralized database
2. A decentralized database
3. A client database
4. A portable storage medium

With a centralized database structure, the biometric information and data as well as the biometric templates are stored onto one database, or a centralized database, as the name implies. This central database is then further networked to all of the biometric devices. As a result, once the biometric templates are processed at the device level, they are transmitted across the network media into the central database for storage.

Also, if biometric template transaction processing occurs at the central database server level, the server then becomes known as a *biometric recognition server.* A key advantage to this type of structure is that remote identification and/or verification are possible.

In terms of a decentralized database structure, multiple database servers are required to store all of the biometric information and data as well as the biometric templates. However, the actual database servers can reside at different locations, many miles away from one another, or they can all be at the same location.

Alternatively, another type of configuration in this particular structure would be to have the central database divided into separate partitions, and then these partitions would become literally their own biometrics database, which can then be stored at different locations. A prime advantage of this type of configuration is that the network processing time is much quicker, and the overall advantage of decentralized database structure is that of redundancy of the biometric information and data and the biometric templates.

With regard to a client database, this refers to storing the actual biometrics database into the biometrics database itself. This type of configuration is very often used in stand-alone biometric devices, with the prime advantage of very quick processing times for either identification and/or verification types of applications.

However, if there are many biometric devices involved in the configuration, then these devices can then be connected to a central server, which thus forms a client–server network topology. All biometric transaction processing and storage, biometric information and data, and biometric templates occur and are stored at the server level. However, the central server keeps the client biometric devices updated when verification and identification transactions do occur.

Finally, with regard to a portable storage scheme, this simply refers to the fact that all of the biometric information and data and biometric templates are stored onto a device that is actually mobile in nature and by design. A prime example of this is the smart card, which consists of a memory chip where the biometrics database can be stored.

Some key advantages of this type of storage scheme is that the end user is in charge of his or her own smart card, and all that is needed is a smart card reader to read and process the biometric templates that are on the smart card. Thus, the issues related to violations of privacy rights should be greatly mitigated, and the feeling of nonintrusiveness greatly increases as well with the use of a smart card.

TEMPLATE MATCHING

After the data storage subsystem, the next subsystem in the biometrical architectural design is the matching subsystem. This is also a very important subsystem, as this is the component that compares the enrollment templates against the verification templates, for either identification or verification types of scenarios.

Specifically, matching is defined as "the process of comparing a biometric sample against a previously stored template and scoring the level of similarity. Systems then make decisions based upon this score and its relationship (above or below) a predetermined threshold" (*Certified Biometrics Learning System, Module 2, Biometrics System Design and Performance,* © 2010 IEEE, pp. 3–32). It should be noted that the matching subsystem can search an entire database, or winnowing techniques can also be utilized.

Rather than searching the entire database, winnowing techniques search only a particular subset or smaller partition of the actual database. To use winnowing techniques effectively, the concept of penetration rates become very important to understand. Specifically, the penetration rate can be defined as "a measure of the average number of preselected templates as a fraction of the total number of templates" (*Certified Biometrics Learning System, Module 2, Biometrics System Design and Performance,* © 2010 IEEE, pp. 3–32).

The lower the penetration rate, the better, which simply means that fewer biometric information and data have to be searched through. To help reduce the penetration rate and make winnowing as effective as possible in searches and retrievals, the concept of binning is also used. The primary goal of binning is to create and develop a much smaller search space to help increase the accuracy rate.

With the techniques of binning, permanent categories on the biometric information and data are created, and these categories are as follows:

1. *Exogenous data*: These are data that are related to the biometric template but are not necessarily derived from it.
2. *Endogenous data*: These are data that are directly derived from the biometric template (such as the minutiae from the fingerprint).

Binning, with the use of preselection algorithms, can either filter or index through the smaller partitions of the biometric information and data. Specifically, a preselection algorithm can be defined as "an algorithm to reduce the number of templates that need to be matched in an identification search of the enrollment database" (*Certified Biometrics Learning System, Module 2, Biometrics System Design and Performance*, © 2010 IEEE, pp. 3–33).

With filtering, the search is conducted within the entire database, and both the endogenous and the exogenous biometric information and data are used to find the biometric templates whose data exactly match the query that was executed. Filtering can also be performed in stages, where both the queries and the data become more refined and specific in the biometric information and data that are being sought for.

Indexing involves the assignment of a numerical value to each piece of biometric information and data to help decrease the search space and increase the retrieval speed. In the end, only those biometric templates whose index are close to one another are searched for, which subsequently increases the search and retrieval time.

It is the matching subsystem that does the comparing and evaluating between the enrollment and the verification templates. Although it appears quite easy and simple to the end user, the matching subsystem is quite a very complex process because the laws of statistics and probability are used to determine an actual match.

There are three types of matching scenarios that can occur:

1. Verification matching
2. Identification matching
3. Vector matching versus cohort matching

With verification matching, as it was closely examined in Chapter 1, the goal is to confirm the claimed identity of a particular individual. In other words, the question of "Am I whom I claim to be?" is being answered by the biometric system. This is known as 1:1 matching, or as mathematically expressed, a 1:1 relationship. In this scenario, only one enrollment template is compared against only one verification template.

With identification matching, as it was also reviewed in Chapter 1, the goal is to confirm the unclaimed identity of a particular individual. In other words, the question of "Who am I?" is being answered by the biometric system. This is known as one-to-many matching or mathematically expressed as 1:N relationship. In this scenario, the verification template is compared with many (perhaps even the entire database) enrollment templates to ascertain the true identity of the individual. Identification is also known as *biometric searching*.

Also, matching can occur in both vector measurements and cohort matching. With the former, vector scores are computed and calculated and are subsequently further broken down into Euclidean (geometric) distances. If the enrollment and verification templates are deemed to be close enough in terms of this distance by the biometric system, then the individual is authorized to gain access to whatever resources he or she has requested.

With the latter, statistical models are actually created based upon the enrollment templates that were constructed prior. The matching process then involves a determination by the biometric system on whether or not the particular signal could have been generated by that certain statistical model. The score is then calculated using statistical probabilities, and if the score is close enough, then the individual's identity will be confirmed.

Finally, in a verification application, the result is often displayed by the system as a numerical score, and with identification applications, the result will be displayed also as a numerical score. However, instead, the identification result is shown in a range from zero to some higher number, as determined by the system administrator. It is then up to the decision-making subsystem within the biometric system to confirm or deny the identification of the individual in question.

THRESHOLD DECISION MAKING

The next step after the matching component is the decision subsystem. Now that there are enrollment and verification templates that can be matched, the next process is to examine the two templates and evaluate the closeness of similarity, or the degree of dissimilarity between the two, and from there make an actual decision if the individual should be authorized to gain access to the resources that he or she has requested.

Specifically, in terms of biometrics, a decision can be defined as "the resultant action taken (either automated or manual) based on a comparison of a similarity score (or similar measure) and the system's threshold" (*Certified Biometrics Learning System, Module 2, Biometrics System Design and Performance*, © 2010 IEEE, pp. 3–36).

For example, if the comparison result between the enrollment and the verification template is above the threshold level, a match is considered to have happened. However, if the comparison result is below the threshold level, then a nonmatch is declared in verification scenarios. It should be noted that the security threshold is normally established to maximize the statistical probability for a correct match to occur.

Also, at this point in the biometric system, the decision-making component is heavily dependent upon the previous subsystems as previously reviewed. The quality level must be high in these previous subsystems for correct matches to occur.

In terms of identification scenarios, the entire database is scanned, and then a subset of the database is presented to the system administrator of possible matched candidates. There are three techniques in which a biometric system can utilize to return a subset of the biometrics database, which consists of the list of possible matched candidates. The techniques are as follows:

1. *Threshold-based*: In this technique, every template in the biometrics database is scanned through entirely, and a comparison score is calculated for every candidate who is a potential match. A potential match is determined if this particular comparison score is above the security threshold level. However, if there are no comparison scores that exceed the security threshold level, then the identity of the person cannot be confirmed because it is assumed then that they have never been enrolled in the biometric system.
2. *Rank-based*: With this technique, each and every biometric template is also scanned in the biometrics database, and a subset of possible matches is returned to the system administrator, with ranks ranging from best to worst.

3. *Hybrid method*: This is a combination of the threshold and ranked methods, and in this scenario, the most common application is to compile and return a ranked list of possible candidates whose comparison scores are well above the established security threshold level.

The security threshold upon which the score is compared against is not just solely decided by the system administrator, but rather by a combination of factors. First, there are three different levels at the security threshold, for both identification and verification scenarios. They are as follows:

1. *Static threshold*: This occurs when the same security threshold level is established for all verification and identification transactions.
2. *Dynamic threshold*: This happens when the security threshold is changed because of variances in both the internal and external environment.
3. *Individual threshold*: These are static in nature as well, but can vary in the level of setting for different employee job roles (for example, the C-level executive of an organization will have a different security threshold level than that of a temporary worker).

Second, the variables used in deciding upon the security threshold setting include the following:

1. The quality score (to the percentage of a biometric raw image that is actually deemed to be usable by the biometric system)
2. The summation of other compiled results from different modalities within a multimodal security solution
3. Any additional information and data that exist upon the comparison score

Third, there are two different subset elements of the above variables that help to further define what the exact security threshold level should be at a much more granular level:

1. Lights-on mode versus lights-out mode
2. The system policy that is set forth for the particular biometric technology

With the first subset element, lights-on mode refers to the fact that some type of human intervention is required, along with the decision made by the biometric system to confirm or deny the identity of the individual in question. With the lights-out mode, the biometric system automatically makes the decision itself, without any human intervention.

With regard to the system policy subelement, this decides whether the match result should be ultimately accepted or rejected. With regard to identification scenarios for this particular subelement, this also helps to decide whether a matched candidate should be accepted or rejected due to the sheer volume of potential candidates that can be returned back. Finally, it is the system policy subelement that helps decide as well the characteristics of a biometric template that is desired to fit the system requirements.

ADMINISTRATION DECISION MAKING

After the matching subsystem, the next subsystem to be encountered in any type of biometric system is that of the administrative system. It is at this point in the biometric device in which human intervention is possible, decision making by a human can also happen, and database monitoring can transpire as well.

The administrative subsystem can be as simple as a keypad on the biometric device itself or it can be as complex as an entire separate software application in a biometrics-based, client–server network topology. Despite what the actual interface looks like, there are a number of critical functions the administrative subsystem serves, which are as follows:

1. The communications role in the biometric system
2. Biometric system control and administration
3. Biometric system auditing, reporting, and control
4. Keeping track of the biometric system with different mode controls
5. Biometric system administrative controls

It should also be noted that the administrative subsystem is designed to act as the primary interface to that of the overall security system in general, as well as the actual hardware, as mentioned previously. For example,

1. Biometric devices are rarely just stand-alone devices any more. For instance, they very often networked together with other types of biometric modalities, thus serving a much larger security application. This can include physical access entry, logical access entry, time and attendance, and perhaps even a combination of all three.
2. However, if a biometric device does indeed have to operate as a stand-alone device, all communications with it are still conducted through it at the administrative subsystem.

However, apart from the internal security communications just described, the administrative subsystem is used also to allow for the biometric system to communicate with the other types of security devices that are nonbiometric related in nature, such as CCTV cameras, background checks technology, door alarms, etc.

It is critical at this point that the administrative subsystem be configured to exactly meet the needs of the system requirements of the overall biometric architecture. Some of these configurations include the following:

1. The formatting of the biometric templates (both enrollment and verification), so that they are in compliance with the various standards as also described earlier in this book.
2. The compilation into a report format of all biometric system statistics and logs to the system administrator. Such data can include items such as various error rates that have been encountered over a certain period, the differing types of biometric transactions that have occurred, authentication/nonauthentication of end users, etc.
3. The reformatting and transmission of biometric templates that are to be sent to other biometric devices, such as those in a multimodal security solution.

The administrative subsystem provides the system administrator the authority to control all aspects of the biometric system, whether it is a stand-alone device or a multimodal biometric security system. However, there are a number of key controls that are deemed to be the most critical:

1. Biometric templates adaptation
2. The establishment of the security threshold values
3. Reporting and control
4. System mode adjustment
5. Granting of privileges to end users

Biometric Templates Adaptation

With the biometric template adaptation control, this merely refers to the fact how often the end user needs to re-enroll his or her biometric templates, or update them, because the templates can degrade over time. It should be noted that it is not the biometric template in the database that erodes over time; rather, template aging refers to the fact that the end user's physiological and biological features can degrade over time, thus causing a mismatch between the newly created verification template that has just been presented versus the older enrollment template that has been captured and created at an earlier time.

Specifically, template aging can be defined as "the increasing dissimilarity between a biometric reference and the actual biometric trait of the user due to periodic changes" (*Certified Biometrics Learning System, Module 2, Biometrics System Design and Performance*, © 2010 IEEE, pp. 3–43).

There are two methods to alleviate the problem of template aging:

1. *Re-enrollment*: If there is a huge difference between the enrollment template and the newly created verification template, all that is required is for the end user to present his or her particular biometric trait so that a new enrollment template can be created.
2. *Template adaptation*: If there is not too much of a difference between the two templates, then the enrollment template is merely *dynamically updated* (or reconstructed) so that the templates (the enrollment and verification) are more closely aligned with one another.

Establishment of the Security Threshold Values

With regard to security threshold establishment control, this refers to the security threshold values at which the end user will either be ultimately accepted or rejected by the biometric system. The exact threshold can be set at a static, dynamic, or individual threshold level. This is probably one of the most difficult decisions to make after the actual biometric system has been installed, and the system administrator needs to take this into consideration.

Reporting and Control

With regard to the reporting and control mechanism, this functionality provides the system administrator the ability to configure and prepare reports with regard to the biometric system in general. Some of the options that the system administrator has at their disposal include

1. The specific data that need to be recorded
2. How long the data need to be archived
3. How often (the exact frequency) the reports need to be published
4. Configuring the reporting tool(s) so that the relevant and appropriate data are presented to the system administrator

System Mode Adjustment

In terms of the system mode adjustment control, there are three separate and unique modes in which the system administrator can set:

1. *Training mode*: This occurs when the biometric system is used to teach the end user about biometrics, how the biometric system works, and how to enroll their respective templates properly.
2. *Testing mode*: This happens when technological (including both hardware and software) upgrades are performed on the biometric system.
3. *Operating mode*: This happens when the biometric system is functioning normally, as required.

Privileges to End Users

Finally, in terms of privilege control, with the administrative subsystem, there are two types of settings that can be established:

1. Access control and security settings
2. Biometric information and data privacy settings

DATA TRANSMISSION

The final subsystem in the biometric architecture model is that of the data transmission subset component. With this subsystem, all of the other subsystems just discussed are networked together and thus bring the biometric system into one cohesive working technology. However, it should be noted that this final subsystem is a rather broad one, and the networking involved includes all hard-wired and wireless networking.

The data that are transmitted in this subsystem primarily include the following:

1. Biometric raw images
2. Enrollment and verification templates
3. Other types of data, which include the authorization/nonauthorization decisions made by the biometric system with regard to the end user

Like the traditional OSI model, the data transmission subsystem performs the encoding of the biometric templates and other biometric information and data as they leave the biometric system and the decoding of the biometric templates and other biometric information and data as they enter the biometric system.

One of the biggest weaknesses of a biometric system lies within the transmission information and data that transpire. One of the biggest threats posed to the data transmission subsystem is that of channel effects. Specifically, it can be defined as "the changes imposed on the presented signal in the transduction and transmission process due to the sampling, noise and frequency response characteristics of the sensor and transmission channel" (*Certified Biometrics Learning System, Module 2, Biometrics System Design and Performance*, © 2010 IEEE, pp. 3–46).

Currently, such tools as encryption and digital signatures can be used to combat this threat. Another emerging tool is that of biocryptography, which will be reviewed completely in Chapter 4.

The section Biometrics Project Management extensively reviewed the system architecture that is involved in a biometric system. This architecture is technology neutral, meaning the system architecture model is applicable to all types of biometric technologies, whether it is physical- or behavioral-based.

BIOMETRICS PROJECT MANAGEMENT

Every biometric device, whether it is a small single sign-on solution that fits onto a netbook or something as large as a hand geometry recognition device, needs all of these subsystems to function and to perform the necessary verification and/or identification transactions.

However, this biometric system architecture model is often viewed as more of a theoretical one to the C-level executive, especially to the CIO and the CFO. What matters to these top-level executives is the actual project implementation of the biometric system after it has been procured and financed for. In other words, before the biometric system can operate in full mode, some key questions need to be answered:

1. How will the biometric system be installed at the place of business?
2. What kind of after implementation testing will take place to ensure that all of the security requirements have been met by the biometric system?
3. More importantly, how will the end users be trained to effectively use the biometric system?

All of these questions and many more have to be answered before the biometric system can go into a live and fully functional mode at the place of business or organization. These questions can only be answered by examining the scope of the biometric system through project management techniques. As a result, this section is intended to provide the CIO, the CFO, and any other C-level executive involved in a biometric system procurement decision the basic project management tools they will need to ensure a successful and effective, long-term deployment.

Specifically, this section focuses on the following project management techniques:

1. Biometric system concepts and classification schemes
2. Biometric system design and interoperability factors and considerations
3. Conducting a biometric systems requirements analysis
4. Biometric system specifications
5. Various biometric system architectural and processing designs
6. Biometric subsystem analysis and design
7. Biometric subsystem implementation and testing
8. Biometric system deployment and integration
9. Biometric system maintenance

System Concepts and Classification Schemes

Long before the biometric modality and the respective hardware and software are decided upon, most C-level executives must first have a clear picture of what they truly want out of a biometric system at their place of business or organization. True, a C-level executive would like to have a biometric system that will provide an enhanced or even a brand new layer of security, but it is also important to consider the other aspects that a biometric system can offer.

For instance, apart from providing physical and/or logical access, will time and attendance for employees also be a desired application as well? These questions and much more must be carefully considered and answered by launching four project management efforts:

1. Conducting an exhaustive system design process
2. Analyzing the system requirements with the appropriate development management life cycle
3. Defining the specific goals of the biometric system (also referred to as the system concepts)
4. Deciding for what application(s) the biometric system will be used for, specifically if it will be used for physical access entry, logical access entry, or time and attendance, or perhaps even a combination of all these
5. Also deciding what specific mode the biometric system will operate in, for example, verification, identification, or even both?

It is important that all of these questions be answered and that nothing is left to haphazard guesswork or even blind assumptions. The framework for this can be answered by using what is known as the *waterfall model*, which is a very popular management methodology to utilize in technical- and security-based implementations. This model is depicted in the following figure.

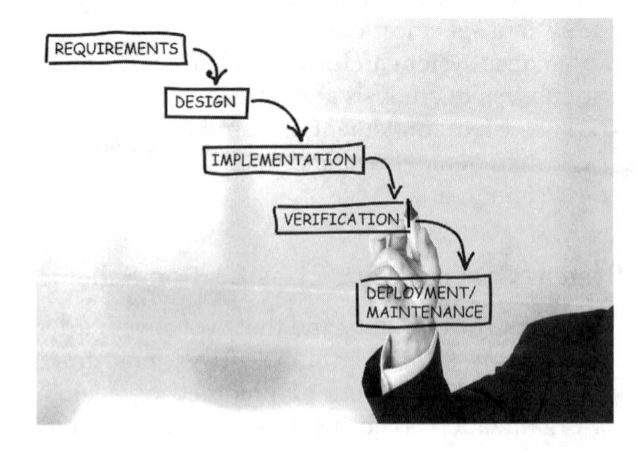

In order to first develop the systems concept and classification scheme component as previously mentioned, four individual processes must be accomplished:

1. Acquire
2. Supply
3. Develop
4. Operations and maintenance

In the acquire process, the C-level executive performs the various RFIs (requests for information) and RFPs (requests for proposals) and prepares the other types of procurement documents that are necessary to help decide which biometric technology will be the best for their place of business or organization.

In terms of the supply process, this is where the biometric vendor(s) comes in. For example, the biometric vendor(s) can present their technology to the C-level executive in the previous process.

In the develop process, the business or organization conducts the actual biometric system analysis and design, the system specifications, the biometric architectural design, and the testing of the actual biometric system before it goes live.

Finally, with the operations and the maintenance process, the biometric system is used for its intended purposes and applications, and must also be maintained as well, which includes the monitoring of the exact performance of the biometric system, and if deemed to be necessary, even migrating to a newer biometric technology and system, thus retiring the existing biometric system.

Upgrading to a Newer Biometric System?

It should be remembered by the C-level executive that upgrading to a newer biometric system can be a very complex and expensive process and undertaking. One of the biggest issues for the C-level executive at this point is whether or not to retain the original biometric templates as well as their related information and data.

The issues faced lie within the fear of the potential abuses of the biometric templates, at least in the eyes of the end users. However, as always, the broader scope and the picture must always be kept in mind, and therefore, by retaining the original biometric templates and their corresponding information and data, the transfer to a newer biometric technology and system will be much easier.

At the systems concept phase, a very high initial biometric systems design is created primarily based upon the needs, objectives, and overall security mission of the place of business or organization. It is very important for the C-level executive to keep in mind that a biometric system is designed to serve a much bigger and broader security purpose.

The systems concepts phase includes the following components:

1. Security problem/needs identification, and based on that, the establishment of the high-level objectives
2. Conducting and analyzing a feasibility study
3. Creating and developing a high-level case study of the intended uses of the biometric system
4. Defining the conceptual architecture of the proposed biometric system

With regard to the security need/problem identification, to gain a clearer picture of the specific security needs of the place of business or organization, the following questions need to be asked:

1. What is the exact security need that needs to be identified?
2. Who will the end users be, and also, what are their exact, specific needs?
3. Who will be the system administrators of the proposed biometric system?
4. What are the restraining variables and factors that must be considered before the biometric system is deployed?
5. What level of performance is required by the proposed biometric system?

Although the chief reason for the procurement of a biometric system is identity management, the needs can vary greatly under this broad umbrella of security. Once the above questions are actually answered, the goals for the proposed biometric system can thus be defined. The C-level executive must also remember that the goals for the proposed biometric system are at a strictly high level and have to correlate strongly with the objectives of the place of business or organization, as defined previously.

The Feasibility Study

In terms of the feasibility study, this involves primarily conducting a cost analysis of the proposed biometric system. The following steps are included with it:

1. Getting a broad understanding of the existing security system at the place of business or organization
2. The considerations of the alternative security solutions and costs in lieu of the proposed biometric system
3. To what degree the existing security infrastructure can be leveraged, and how the proposed biometric system can fit into it
4. The identification of the financial risk areas
5. The development of the proposed budget and the financial constraints that go with it

The C-level executive should also keep in mind that after the feasibility study is conducted, the benefits must be greater than the costs associated with the proposed biometric system for the place of business or organization to realize a positive return on investment (ROI).

Application Classifiers

Application classifiers are used to help further ascertain and drill down the biometric system and design. Classifiers are used in project management extensively today to help ensure a successful deployment, whatever it may be. In terms of biometrics technology, the most commonly used classifiers are as follows:

1. Cooperative users versus uncooperative users: A cooperative user is one who is actively wanting to be confirmed by the biometric system to gain access to the confidential resources and files that they are seeking. This even includes impostors who are trying to break through the defenses of the biometric system. An uncooperative user is one who does not wish to be either identified or identified by the biometric system.
2. An overt biometric system versus a covert biometric system: An overt biometric system simply means that the end user is fully aware and cognizant that his or her physiological and biological traits are going to be captured by the biometric system and that full cooperation is required on his or her part. A covert biometric system is where the end user is not at all aware about the biometric system and that his or her physiological or biological traits are being collected by the biometric system. As a result, the end user is neither cooperative or noncooperative in this certain scenario.
3. Habituated users versus nonhabituated users: A habituated user group is one who is expected to use the biometric system on a

regular basis, without too much resistance or opposition. This also means that this type of user group can undergo a complex and in-depth training program if need be about the biometric system they are using. With a nonhabituated user group, this means that the end users are expected to use the biometric system very infrequently and with much opposition. As a result, the biometric system must be very easy to use, in order to accommodate the very low learning curve of this particular end user group.

4. An attended biometric system versus an unattended biometric system: With an attended biometric system, human intervention is always required on a daily basis, especially to accommodate the uncooperative end user group. The enrollment and verification processes must always be supervised. With an unattended biometric system, no human intervention is required.
5. A constrained biometric system versus an unconstrained biometric system: With a constrained biometric system, the external environmental conditions remain identical on a daily basis for enrollment and verification of the end users. These conditions include the illumination, the type of sensor used in the biometric system, conditions in both the external and the internal environments, etc. With an unconstrained biometric system, there are variations or differences daily in the environment in which both enrollment and verification occur in the biometric system.
6. Private users versus public users: Private users are those that have a vested use and are required to use the biometric system, such as employees of a place of business or organization. Public users are those that are not required to use a biometric system, and if they use it, it is very infrequent.
7. An isolated biometric system versus an integrated biometric system: With an isolated biometric system, the biometric templates and the corresponding information and data are never shared with other security systems. With an open or integrated biometric system, the biometric templates and the corresponding information and data are shared with the other security systems, whether it is biometric-based or not.
8. A scalable biometric system versus a nonscalable biometric system: A scalable biometric system is one that can be upgraded or even replaced with newer technology, hardware, and software,

in order to meet the increased demand. A nonscalable biometric system is one that works best for a fixed number of an end user population and cannot be upgraded or replaced if demand increases over time.

With regard to the above biometric application classifiers, a verification mode is much less taxing on the resources of a biometric system than that of an identification mode. This is so primarily for two reasons:

1. A verification mode only depends upon one single match result. An identification mode relies upon many matches, and the more matches that have to be made to confirm the identity of the individual, the greater the chances for errors and mistakes to happen.
2. A verification mode can be accomplished in just less than 2 or so because only one match needs to be made. However, with an identification mode, since many more matches have to be made, confirming the identity of an individual can take several minutes or more; thus, rendering the biometric system infeasible in extremely quick processing time is an absolute, crucial requirement.

System Design and Interoperability Factors and Considerations

Following the biometric systems and application classifiers, the next project management step is that of the biometrics system design and interoperability factors and considerations. This is the part of the project management process where choosing how to develop the biometric system becomes a key question.

COTS-Based Biometric Systems

For example, a biometric device can be created within the actual place of business or organization (assuming that such expertise actually exists) or the development and creation of the biometric device can be outsourced to a third party or the biometric device can be bought off the shelf, which is also known as *COTS* (commercial off-the-shelf biometric system).

However, whenever it comes to the decision of either making or buying the biometric device, the cost and the least path of resistance to creation and development become the two driving factors. As a result of this, typically, COTS-based biometric systems are picked because of the much lower purchase price that is involved and the fact that a COTS-based biometric system is literally plug and play—just buy it and install it. It should be noted that COTS-based biometric systems are also very technically advanced and can be ready to serve market applications of physical access entry, logical access entry, and time and attendance, quite literally, out of the box.

In addition, apart from the actual hardware, there are also COTS-based biometric software available, which is also plug and play. The benefit of this is that the cost is much lower than hiring a software development firm to design the biometric software application. When a COTS-based biometric system is acquired, it is often not interoperable with other types of biometric systems.

Interoperability is a must if the COTS-based biometric system is to interact with the other security systems at the place of business or organization or if the C-level executive decides to create a multimodal biometric solution. There are two types of interoperability that are available for biometric systems:

1. The ability for other security systems or other biometric devices to be able to work with the COTS-based biometric system in full harmony
2. The ability for all of the biometric devices and nonbiometric devices to exchange biometric information and data readily

The C-level executive should keep in mind that if the COTS-based biometric system is not interoperable, a phenomenon known as *vendor lock-in* can occur. This happens when a COTS-based biometric system cannot just be simply upgraded to a newer system, or rather, the entire biometric system must be replaced to modernize it with the latest technology.

If this does indeed happen, the costs and expense can rise exponentially. However, if the COTS-based biometric system is designed and made to be interoperable, then simple upgrading is a much less expensive burden for the C-level executive. To help with the issue of interoperability, standards can be developed to help biometric systems become more integrated with one another, interchanged with one another, and even upgraded over time.

Proprietary or Open-Ended System?

However, standards can only go so far in helping with biometric system interoperability. Another key piece of the puzzle is the choice of whether to deploy a proprietary biometric system or an open-ended biometric system. With a proprietary biometric system, the only choice that a C-level executive has is that if he or she wants to upgrade his or her current biometric system, the existing one will have to be literally ripped out and replaced with an entirely brand new system. This means great expense will be incurred by the place of business or organization, but an open-ended biometric system will have a much higher level of interoperability than a closed system. Thus, if a C-level executive wants to pursue an upgrade, all that will be required is a replacement of the hardware and software, which will only be a fraction of the cost of installing an entirely brand new biometric system. Also, upgrading a biometric system is much easier if open-source software is used and follows open-source biometric standards such as BioAPI.

Infrastructure Assessment

Another important aspect of a biometrics system and interoperability factors and considerations is that of conducting an assessment of the infrastructure in which the new biometric system will be deployed. The infrastructure assessment involves a careful analysis of the following:

1. Physical infrastructure
2. Platform infrastructure
3. Network and directory infrastructure
4. Database infrastructure

With the physical infrastructure, this aspect must be considered very early on. This part examines the physical premises, power capacity, the blueprint layout, as well as any types of enhancements that need to be made to accommodate the specific biometric system that will be deployed.

For example, special lighting may be required for a facial recognition system, or there could be different mounting heights that need to be considered for a fingerprint recognition device.

In terms of the platform infrastructure, this is also a crucial support that needs to be considered very seriously. For example, the chosen platform needs to be able to provide enough processing power (and of course, the more the better) to meet the high demands and speed of the biometric system, especially when it comes to verification and identification transactions.

The chosen platform must also be compatible with the hardware and the software that are being supplied by the biometrics vendor.

With regard to the network and directory infrastructure, it is also very crucial to have superb bandwidth that will provide enough capacity and throughput for the quick transmission of the biometric templates. The C-level executive must remember that the point from where the biometric raw image is captured will be different from the point at which the verification and identification transactions and decision making occurs (this will most likely occur in a separate server).

Thus, there can be no bottlenecks whatsoever in the network if verification transactions are supposed to occur in less than 2. Also, if the biometric system is integrated or coupled into a directory structure (such as Microsoft Active Directory), the choice of communications protocols to be used also becomes important.

Finally, the database infrastructure is where the biometric templates will be stored; thus, the choice of software and the design of the biometric system are of utmost importance. Also, key questions such as who will own and manage the database and especially how the biometric templates will be protected must be answered before the biometric system is procured.

The Human Equation

In this phase of the biometrics project management, the views of the people who are involved and impacted by the biometric system need to be taken into serious consideration as well. In the end, it is not just the C-level executive who will be affected by the deployment of a new biometric system but also a whole group of people.

This group of people is known as the stakeholders and include the following:

1. The individuals who are responsible for the procurement and the acquisition of the biometric system (for example, the C-level executives and other members of middle management)
2. The individuals who will monitor and run the biometric system (for example, the end users and the IT support team)
3. The individuals who will be using the biometric system (for example, the employees of the place of business or organization and other visitors)

It is important that during each phase of the biometric project that all of the stakeholders are involved in the entire process and not to exclude any one group of individuals. That way, by being involved in the entire process, the acceptance level of the stakeholders for the biometric system will be much higher in the end.

Given the very important role of the stakeholders in the deployment of the biometric system, we now turn our attention to the more psychological aspects that need to be taken into consideration as well. Biometrics is one of those security technologies that, when placed upon a spectrum with the other technologies, has the most social impact upon end users and the public at large.

Therefore, it is of crucial importance for the C-level executive to take into consideration all of the psychological aspects of the biometric systems and the impacts they will potentially have upon all of the stakeholders. Although all of the psychological issues may not be resolved, if they are at least addressed from the outset, this will help to bolster end user acceptance in the end.

Some of the issues, both psychological and physical, that end users experience with biometric systems include the following:

1. Very different viewpoints and differences as to the exact human reactions to the concept of biometrics in general
2. The varying levels of acceptance of using biometrics technology
3. Any type of medical conditions or illnesses that could affect and impact both the physiological and the behavioral traits of the end users
4. Manual labor jobs, which can affect the quality of both the physiological and behavioral traits (such as construction workers having their fingers impacted because of their job requirements)

Ergonomic Issues

A key component of the psychological issues of biometric systems is that of ergonomics. As it specifically relates to biometrics, ergonomics takes into consideration how to make a biometric system more comfortable for all of the end users to enroll their physiological and behavioral templates.

Some of these considerations include the following:

1. Height variances and differences among the end users
2. Frequency of usage of the biometric system
3. Design of the physical space of the biometric system by giving the end users a sense of privacy

Another key aspect in addressing the psychological aspects of a biometric system is to foster open and honest lines of communication with the end users, especially the employees of a place of business or organization. By allowing such a line of communications to be established will only help to foster trust and acceptance of the biometric system by the end user.

The Feedback System

In this regard, feedback to the end user becomes very important. This can also be referred to as a feedback system. It can be represented by the following mechanisms:

1. Audio feedback in different languages to accommodate end users of different cultures and different countries
2. Process pictures depicting to the end user how to easily enroll their physiological and behavioral traits into the biometric system
3. Any type of on-screen graphical user interfaces (GUIs), which are easy to comprehend and understand
4. Other visual cues (also known as *passive feedback*) to help guide the end users through the enrollment and verification processes

It is best to accommodate the end users by offering a combination of the above feedback mechanisms. In this regard, as a result, all types of end users can thus ultimately be served. Also, deploying a biometric technology can bring up fears and anxiety to the end users because it is such a huge change.

This is where the appropriate change management mechanisms can play an important role as well. Specifically, change management can be defined as "the process of preparing the organization for a new system" (*Certified Biometrics Learning System, Module 2, Biometrics System Design and Performance*, © 2010 IEEE, pp. 3–64).

To make change management easy and effective and help ensure a smooth transition, clear and open communications from upper management to the end users is extremely vital.

It is thus important that such communication starts at a very early phase, perhaps even when the thought process of the consideration of a biometric system for the place of business or organization comes to the minds of the C-level executives. In this aspect, communications can take place in the following forms:

1. The advertising of the benefits of the new biometric system to the end users
2. Addressing the privacy rights and civil liberty fears and issues the end users may have about the new and upcoming biometric system
3. Keeping the end users informed that enrollment into the biometric system is not mandatory and that other security options are available
4. Preannouncement of formal training sessions into the proper and effective use of the biometric system

Because the biometric system has such strong impacts on the end user, its acceptance and the end user's ability to use a biometric system can actually have a much more pronounced, positive effect on system performance. Conversely, if the biometric system is poorly received by the end user of the place of business or organization, then the system performance could quickly deteriorate in just a short matter of time.

However, if the place of business or organization is very large, for example, with thousands of employees located in offices all over the world, then a study called *population dynamics* can be utilized to study the entire end user base to create a more effective use of the biometric system.

Population Dynamics

Population dynamics is actually a statistical analysis that examines the average end user. Within population dynamics, two types of variability can be demonstrated:

1. *Interclass variability*: This is the amount of uniqueness, or the degree of distinctiveness, among each end user. A much lower interclass variability means that there could be potentially escalated FARs (meaning an impostor is accepted and authorized by the biometric system) or even higher FERs (this happens when the end user simply cannot enroll into the biometric system).
2. *Intraclass variability*: This is the amount of uniqueness or distinctiveness between repeat measurements of the same end user. If this particular variability is high, there could be an escalated rate of false nonmatches occurring during the verification or identification process.

Businesses or organizations that are very large (for example, over 5000 employees located globally) will have a large-scale biometric system implemented as well as large-scale biometric software applications.

Many C-level executives at this point question what is the exact criterion that defines a large-scale biometric implementation at this phase in the biometric project management life cycle. Although there is no hard and fast rule, most large-scale biometric systems and applications start at the range of 100,000 templates and greater.

However, not all biometric technologies are well suited for such large applications. There are two key variables that can pose a huge challenge when businesses and organizations try to implement large-scale biometric implementations:

1. Accuracy
2. Speed

For instance, there is a negative correlation that exists between the two with regard to the scalability of a biometric system, i.e., as biometric systems and applications get larger in size, in terms of the total number of biometric templates that are stored and processed, the accuracy and speed decline at the same, if not greater, rate.

Thus, the question now is, which of the biometric modalities are best suited for large-scale applications? Here are some key points for the C-level executives to consider as they evaluate upward expansion of their biometric system(s):

1. Utilize a biometric system that typically has a low FMR, as this tends to scale up naturally to much larger biometric databases.
2. Implement a multimodal biometric system to help assure large-scale biometric database scalability.
3. Utilize the techniques of winnowing, such as filtering, binning, and indexing, to help with upward biometric system scalability.
4. Move the existing biometric templates to the newer, much larger biometrics database, rather than have the end users re-enroll their physiological or behavioral traits. The end result of this is that this will save a lot of expense and time when scaling upward.
5. Figure out exactly the network throughput that is required when the biometrics database is scaled upward.
6. Calculate the difficulty in scaling upward the existing biometric system to the hypothetical largest biometric system that con-

ceivably the place of business or organization could potentially support.

7. Try to estimate the per unit cost of the biometrics database processing versus the total cost of processing the transactions that are needed for the entire end user population of the place of business or organization that is using the biometric system.

Systems Requirements Analysis

The next step in the biometrics project management life cycle is that of the biometric systems requirements analysis. Determining the requirements of what is demanded by a potential security system is one of the most important aspects of the biometrics project management life cycle. For example, defining the specific requirements can be used to further pinpoint the security problems that the place of business or organization is experiencing and the goals that can be used to help allay those specific problems and risks.

The term *requirement* is a broad one, but for the purposes of biometric system deployment, it can be specifically defined as follows: "a condition or capability needed by a user to solve a problem or achieve an objective" (*Certified Biometrics Learning System, Module 2, Biometrics System Design and Performance*, © 2010 IEEE, pp. 3–69).

In biometrics project management, defining a set of requirements aids in the process of

1. Clearly identifying absolute needs from unclear wants
2. Ascertaining the specific functions and specifications of the new, potential biometric system
3. Determining what the GUI will be, and how it will appear in terms of looks to the biometrics system administrator
4. Providing a review of how the biometrics database will be designed, how it will function, and how it will integrate into the biometric system hardware and software
5. Establishing the guidelines and the specific criterion for biometric system testing after it has been deployed and before it is made available to the end user in the place of business or organization
6. Creating the biometric system documentation for maintenance, and more importantly, for end user training and use

The biometrics system requirements consist of the following steps, and it will be reviewed in detail:

1. Biometric system requirements elicitation
2. Biometric system requirement analysis and regulation
3. Biometric system requirements documentation
4. Biometric system requirements validation
5. Biometrics requirements management

System Requirements Elicitation

With regard to the biometric system requirements elicitation, this is where all of the stakeholders who are involved in the biometrics project list out their specific needs and concerns about the potential biometric implementation. As a result of this process, end user acceptance of the biometric system should dramatically increase once the system is deployed because the stakeholders will feel that they are taking an important part in the entire process.

Also, this process helps the technical staff (especially the IT staff) to pick up a biometric system that is specific to the security needs of the place of business or organization. It should be noted at this point that the biometrics requirements elicitation can either be passively or actively based. With the former, focus group interviews can be conducted with all of the stakeholders who are involved, and with the latter, process diagrams and other types of technical documentation are created.

Another effective tool that is utilized in this process is that of use cases. These are actually process flow diagrams that are used to map the biometric system in its entirety at the place of business or organization. To aid in this particular diagramming process, very often, the unified modeling language, or UML, is widely utilized.

System Requirement Analysis and Regulation

In the second phase, which is biometric system requirements and negotiation, several steps are involved, which are as follows:

1. Rank requirements
2. Model requirements
3. Ascertaining if the potential, new biometric system can actually solve the security needs of the business or organization

With rank requirements, the IT staff and the system administrators try to ascertain the statistical probability of severity if a particular system requirement were not to be implemented for any type of reason.

If the absence of a requirement causes a severe problem to the business or organization, then it is ranked with a very high statistical probability. However, if the absence of a requirement causes a less severe problem to the business or organization, then it is ranked at a much lower statistical probability. Also, cost functions associated with the biometric system can also be ranked on a more severe or less severe statistical probability rating as well.

With regard to the model requirements, as described previously, modeling languages are utilized to map out the potential, new biometric system. To assist with this task, two types of modeling languages can be utilized:

1. *Unified Modeling Language (UML)*: As described previously, this is a visual design software package that is used to graphically design the various components of the entire biometric system.
2. *Biometric and Token Technology Application Modeling Language (BANTAM)*: This is another visual design package in which the hardware and software aspects of the biometric system can be mapped out quite easily.

Finally, ascertaining if the biometric system can meet or fulfill the security requirements can help the IT staff, system administrators, and the C-level executives of a business or organization determine if their security needs can be met with just a biometric solution or if another biometric or even nonbiometric technology is needed to create a multimodal security solution to fulfill the specific security requirements.

System Requirements Documentation

With biometric requirements documentation, this is where all of the steps and components of the proposed biometric system are documented and tracked. After the biometric system is implemented, this documentation then serves as the focal point for troubleshooting and creating end user training booklets and guides.

In terms of biometric system requirements validation, this is the process where all of the stakeholders who are involved in the implementation review the documentation created in the previous step to find any errors or points of confusion in order to ensure that the biometric system implementation goes as smoothly as possible and that all foreseeable problems are resolved ahead of time.

Also, prototypes of the actual biometric system can be obtained from the vendor and a test environment can be simulated to further ensure that the real biometric system implementation will go smoothly and just as planned. Once all of these have been accomplished, all of the stakeholders sign off to begin the deployment of the biometric system.

System Requirements Validation

In the last step, the biometric requirement management phase, new documentation is created and updated so that it remains valid if upgrades and/or changes occur in the newly deployed biometric system. Also, during this entire process just described, the place of business or organization can put out RFPs to various biometric vendors to help narrow down which biometric systems will fit into the budget and meet the security requirements.

Also, during this timeframe, the biometric vendors who make the short list can also provide to the business or organization trial products to closely evaluate which biometric system will meet the security needs the best.

Biometric System Specifications

After the biometric system requirements analysis phase, the next phase in the biometric project management life cycle is that of the biometrics system specification phase. This part of the project management phase provides for what the biometric system will accomplish and what it will not. This phase helps the C-level executives keep the biometric system costs in check and keep the implementation time running as scheduled.

The biometric system specification phase consists of the following components:

1. *Biometric system functionality*: These are the specific security tasks that the biometric system is supposed to accomplish.
2. *Biometric system external interfaces*: These are the specifications the biometric system will interact with, such as software applications and other related hardware applications.

3. *Biometric system performance*: These are the specifications as to how each subcomponent and how the entire biometric system as a whole is expected to perform.
4. *Biometric attributes*: These are primarily the specifications for the biometric system maintenance.
5. *Biometric system design constraints*: These are the best practices and standards that the newly implemented biometric is supposed to follow.
6. *Biometric system testing*: These specifications layout the criteria that will confirm what is acceptable performance for the biometric system.

From these above components, the biometric system specifications can be written internally either by the project management team from within the business or organization or by a neutral, unbiased third party. However, regardless of who is composing and writing the documentation, the biometric system specification document will consist of the following components:

1. *The introduction*: This portion of the document consists of the executive summary, the scope of the biometric system deployment and implementation, the key terminology used in the process, and the references used for research conducted to justify the need for a biometric system.
2. *Product description*: This portion describes the entire biometric technology that is going to be used at the place of business or organization and how it will be deployed, such as
   a. The total number of biometric devices to be deployed both outside and inside as well as the total number of end users expected to use each biometric device
   b. The members of the biometric system administration and support staff
   c. The envisioned biometric system networking environment (which will be described in detail in the next section)
   d. How training will be conducted for the end users
   e. How the privacy of the end user's biometric information and data will be enforced
3. *System requirements*: This final portion of the document lays out the details of the formal security requirements for the biometric system to be procured and deployed.

System Architectural and Processing Designs

The next phase in the biometrics project management phase is that of the biometric system architectural and processing design. This is the blueprint that is utilized to specify the very exact biometric hardware and software that will be deployed at the place of business or organization. In terms of biometrics, the architectural design component consists of three primary objectives:

1. *High-level goals*: This describes the final biometric system security goals.
2. *Mid-level objectives*: This describes where the biometric system will be placed and how the biometric templates will be matched with and compared with one another.
3. *Detailed high-level objectives*: This helps to define which specific hardware and software will be used in the biometric system.

In other words, whatever the previous phase (the systems specifications phase) examined, it is all now decided upon and finalized by the stakeholders in this phase. This phase consists of two primary objectives—namely, the functional architecture and the operational architecture. These two components also consist of subcomponents and are described as follows:

1. *Functional architecture*: This component describes the actual physical aspect of the biometric system hardware, and the type of environment in which it will be placed at the business or organization. It addresses the following subcomponents:
    a. *Environment design*: This describes how the ergonomic environment will be designed for the biometric system.
    b. *Hardware design*: The specific design of the hardware of the biometric system is a direct function of the exact purpose of the biometric system and the type of environment in which it will be placed. The hardware design of the biometric system is also strongly correlated with the operational architecture, which is discussed next. It is also important to note that the biometric system hardware design is also strongly correlated to the existing security requirements of the business or organization as well as the existing security infrastructure. As a result, the proposed biometric system may be able to integrate with legacy security systems, thus providing for multiple layers of security defenses.

2. *Operational architecture*: This type of architecture lays out the design of the software system of the entire biometric system, and this is what enables the functional architecture to do its part of the work as well. It should be noted that this design of the software system is only a high level. It does not go into the granular level of detail as to how the exact software code should be written, but there will be enough detail in it so that the software developers who are involved will know at a cursory glance what exactly needs to be developed. Within this architecture component, the goals of determining how the biometric templates will be stored and where the matching algorithms will be stored are also determined. In terms of multimodal biometric solutions, the operational architecture also specifies as to how the different biometric modalities will work together.

Storage and Matching Combinations

From this, six different types of storage and matching combinations are most commonly used:

1. *Store on server, match on server architecture*: In this type of architecture, the biometric raw images that are collected from the sensors and the subsequent templates that are created are both stored and compared/matched with one another at the server level. This is also one of the most utilized architectures, and the typical application for which this is used are large-scale (1:N) scenarios.
2. *Store on client, match on client architecture*: With this architecture scenario, the biometric raw images and the biometric templates are stored at the client level, and all processing of the data and information are conducted at the local level. This type of architecture can operate and function without a centralized network, and the most common application for this particular architecture is that of single sign solutions for smartphones and laptops.
3. *Store on device, match on device architecture*: In terms of this operational architecture, the biometric raw images and the biometric raw templates are stored upon the actual authentication device itself, and again, processing of information of the data and matching of the biometric templates take place at the local level. The most common application for this is for verification scenarios on wireless devices.

4. *Store on token, match on token architecture*: Under this operational architecture regime, the biometric raw images and the corresponding biometric templates are stored on some of the physical tokens, such as a smart card. The processing of the biometric templates takes place over a wireless transmission between the smart card and the smart card reader. Because the information and the data are stored on the smart card, different biometric modality templates can be stored upon it as well. The most common application for this particular architecture is that of credential-based security systems, such as turnstiles for physical access entry at large office buildings.
5. *Store on token, match on device architecture*: In regard to this type of operational architecture, and just like the above-described architecture, the biometric raw images and the corresponding templates are stored onto the memory chip of the smart card. However, the primary difference is that the matching and the processing take place on the smart card itself, without the need for a wireless connection and a smart card reader. One of the most common applications for this type of architecture is that of worker-based credentialing systems.
6. *Store on token, match on server architecture*: With this type of operational architecture, the biometric raw image and the corresponding templates are stored onto a smart card, but the processing and matching of the biometric templates take place at the server level. A primary advantage of this is that there is no central storage of the biometric templates involved, and all decision making occurs in a very secure environment. A very common application for this kind of operational architecture is that of network and Internet access.

If one were to actually compare all of the subsystem locations just reviewed, a cross comparison of these scenarios can now be made. For example, storing biometric templates at a server level could prove to be much more complex and much more financially expensive versus a local storage device.

For instance, if the server level is utilized, there will be costs associated with licensing, hardware and software upgrades, as well as hiring a full time server administrator. Therefore, the use of a server-based system would probably work best for large-scale identification-based applications.

For verification-based applications, local storage, such as the level of a biometric device or a smart card, possesses some unique advantages:

1. Other types of wireless readers can be used (such as for the e-passport applications).
2. No need to have the costs associated with a centralized server and database.
3. Multipurpose biometric applications can be much easier deployed and served to the end user base.

Operational Architecture Design—Multimodal Systems

The third advantage in the preceding list leads into the operational architecture design for multimodal biometric systems. It is important to keep in mind that multimodal biometric systems are composed of an entire security package; therefore, different types of biometric information and data will be shared. This is also referred to as fusion, and there are four distinct levels of it:

1. *Sensor-level fusion*: At this particular level, biometric information is the most unique and rich but also contains a lot of extraneous background noise that could greatly affect the template creation process.
2. *Feature-level fusion*: Biometric information and data at this level are much less unique than the previous level, but the extraneous noise is much less. There are also two different subfusion layers at this particular level:
   a. *Unimodal feature fusion*: Single biometric information and data fusion can be applied.
   b. *Feature normalization*: A biometric system with different modalities may require normalization, which extracts unique features based upon statistical means and variances.
3. *Score-level fusion*: At this stage, biometric information and data are extremely simplified, and this is a very common method that is currently used because of its ease of use.
4. *Decision-level fusion*: At this particular level, the biometric information and data are very much limited, and this technique is only used when no other biometric information and data are available. This is also referred to as rank level fusion in large-scale identification-based applications.

A basic principle to be remembered here is that when there are multiple sources of biometric information and data that are available, this is much better in terms of unique feature extraction rather than just relying upon one source of biometric information and data.

The Information Processing Architecture

Within this also lies what is known as the information processing architecture. This typically involves the design of any type of computer that is needed as well as the mathematical algorithms and the computer networks needed to support the entire biometric system at the place of business or organization. These are reviewed in the following order:

1. *Computer hardware*: For biometric systems, speed is one of the most important functions. However, it is not only the speed of the biometric system that is important; the speed of the computer is also equally important. After all, as discussed previously, if a large-scale biometrics system is deployed for identification, a centralized computer server is needed not only to store the enrollment and verification templates but also to process all of the transactions that are required. Because all of this requires extra resources, speed will be required on the part of the server processor. In this regard, speed is a direct function of
   a. The generation of the CPU and its speed (measured in hertz)
   b. The arithmetic logic unit that regulates mathematical calculations
   c. The bus speed of the CPU, which is measured in the speed and the width
   d. The input/output rate in which biometric information and data can be sent to other hardware devices that support the entire biometric system
2. *Parallel and distributed processing*: In most large-scale biometric systems, just one processor is not enough. Therefore, the workload must be shared. To accomplish this, parallel and distributed processing are often utilized:
   a. *Parallel processing*: This is defined as "the simultaneous transfer, occurrence, or processing of the individual parts of the whole, such as the bits of a character, using separate facilities for the various parts" (*Certified Biometrics Learning System,*

Module 2, Biometrics System Design and Performance, © 2010 IEEE, pp. 3–90).

In other words, one server processor can send the workload into different components to other server processors to execute a portion of the larger program that supports the biometric software applications from within the system.

b. *Distributed processing*: With this type of processing, the same task is broken among a series of server processors. In this instance, a single mathematical instruction, multiple data architecture is utilized. This simply means that the server processors perform the same, repetitive tasks on different sets of biometric information and data.

The parallel and distributed processing components as described above are a direct function of what is known as *execution efficiency.* It is specifically defined as "the degree to which a system or component performs its designated functions with minimum consumption of time" (*Certified Biometrics Learning System, Module 2, Biometrics System Design and Performance,* © 2010 IEEE, pp. 3–32). This can also be referred to as what is known as *computational expense.*

This is greatly impacted by the sheer complexity of the mathematical algorithms that are involved in a biometric system as well as the sheer amount of biometric information and data that must be processed. As a result, it should be noted that there is a distinct trade-off between the processing quality of the biometric templates and the amount of memory that is required.

Subsystem Analysis and Design

The next phase of the biometrics project management life cycle is that of the biometrics subsystem design phase. In any biometric subsystem, in theory at least, the subsystems are designed to be logically separated from one another. To accomplish this task, the concepts of coupling and cohesion are utilized. Coupling can be defined as "the manner and degree of interdependence between software modules," and cohesion can be defined as "the manner and degree to which the tasks performed by a single software module are related to one another" (*Certified Biometrics Learning System, Module 2, Biometrics System Design and Performance,* © 2010 IEEE, pp. 3–92).

For example, coupling is used and designed in such a way that each subsystem in the overall biometric system can operate separately from one another. To illustrate this point, as the biometric information and data are passed from one subsystem to the next, control over each subsystem is also maintained in terms of execution.

With respect to cohesion, every subsystem in the overall biometric system only performs and executes those specific tasks it has been assigned, to ensure that the biometric system works as one harmonious unit.

In this part of the biometric subsystem, it is important to keep in mind that the biometric system is intended to be what is known as the *human to machine interface*. As a result, the choice of sensor becomes of paramount importance to capture and collect the biometric raw images (in this respect, physical biometrics, primarily).

The design and ultimate choice of a sensor not only has a huge and direct impact upon the performance of the biometric system, but a phenomenon known as *sensor fatigue* can also set in over the useful lifetime of the biometric device. This occurs when the sensor can no longer capture and collect any useful and meaningful biometric raw images.

Also, the environment in which the biometric system will be placed and the type of application it will be used for are also other important considerations that need to be taken into account, and this can have a direct impact upon the choice and design of the sensor as well. The signal and image processing subcomponent of the overall biometric subsystem is dependent upon a number of key factors, which include the following aspects:

1. The specific biometric modality that will be required by the place of business or organization
2. The type of sensor that is needed
3. The preprocessing and the feature requirement of the biometric raw image
4. Any changes of the preprocessing order
5. Any modular designs that can be incorporated
6. The types of the compression ratios that are required for processing the biometric raw images

Data Storage Subsystem Design Considerations

Next in line after the signal and image processing is that of the data storage subsystem design. As discussed before, this is the subsystem that consists of the database that houses the biometric templates and the transaction processing that occurs.

Biometric database design is a very complex task, and at a macrolevel, some of its factors need to be examined closely, such as the overall biometric system requirements and the exact security needs of the business or organization. Also, another key element that helps to determine the biometrics database design is the overall management of the database.

For example, a small database designed for verification (1:1) applications will be much easier to manage than a large-scale database application (1:N). Also, the database designer(s) need to decide if the biometrics database will only contain the enrollment and the verification templates or if other types of biometric information and data will be incorporated into the database or not.

A smaller database of course means less processing power is required, and the search time is greatly enhanced. Another issue involved with the biometrics database design is where to store the biometric raw images. Typically, these raw images are not stored in the biometrics database, but if they are stored for any type of reason, the following factors must be taken into consideration:

1. The biometric raw images should be kept in a separate database to optimize performance.
2. The search time for the biometric raw images can be much slower because they are not accessed frequently.
3. The biometric raw images, if they are stored in the database, will need to be reformatted, so that they can be used to replace the biometric templates that degrade in quality over time or that are accidentally deleted from the system.
4. The biometric raw images need extra security layers, but these protocols should not be applied to the database that houses the enrollment and verification templates, as it will greatly lessen the search and retrieval time.

Database Management System

A biometrics database is normally run, administered, and maintained by a database management system, also known as a *DBMS*. A place of business or organization can either utilize a closed- or open-source database. Both of these have advantages and disadvantages. For example, with the former, although there would be strong vendor support, the software and the respective licensing fees would be cost-prohibitive for a small- to medium-sized business.

However, with the latter DBMS option, the source code is virtually free, but due to its open-source nature, it can be prone to large-scale attacks and other such vulnerabilities. The DBMS for a biometrics database can either be small or large. If it is large, then other considerations need to be taken into account such as the following:

1. The maximum number of end users and their corresponding templates that can be stored into the biometrics database
2. How the biometrics DBMS will handle search and retrieval requests
3. The overall design of the DBMS, specifically the database structure
4. If and when parallel and distributed processing (as discussed previously) will be utilized
5. Specific disaster recovery plans for the DBMS and the corresponding biometrics database structure
6. The frequency in which the biometric information and data in the DBMS will be *cleansed*

The matching subsystem (in which the enrollment and verification templates will be matched against) design will be dictated primarily by the type of biometric modality and the vendor who will be ultimately decided upon by the business or organization. Every system design tries to have a compact matching subsystem component, but this is 100% dependent upon where the matching will take place.

If the matching subsystem will be placed in a client–server network topology, the matching subsystem can be designed as small as possible because the server has dedicated resources that can be used. However, if the matching subsystem is to be placed onto a smart card, then the resources for matching have to be as maximum as possible, given the constrained nature of a smart card.

With regard to the decision subsystem design (this is the part when the final decision is made whether or not to confirm the identity of the person in question and whether or not to grant them access to the resources they are seeking to use), other considerations need to be taken into account as well, such as the following:

1. Whether a supervised or a nonsupervised system will be utilized
2. The point at which positive verification or negative verification (similarly, also positive verification and negative verification) will

be made based upon the comparison of the enrollment and verification templates
3. If only one security threshold will be used or other security thresholds will be used as well to provide a sense of randomness to the biometric system

Determining the Security Threshold

As it was reviewed earlier in this book, deciding upon what the exact security threshold should be can be a very complex process to achieve in order to garner the most optimal result from the place of business or organization. For example,

1. Very often, ascertaining the exact level of the security threshold is first done in a test environment. If the end user database varies from the test database, then the ascertained security threshold setting will be completely off base.
2. A very strong barrier in ascertaining the true, optimal security threshold setting is further exacerbated by the fact that it is statistically very difficult to determine what the true rejection rates are.

Also, two other key factors that are very important in ascertaining the security threshold level are the type of biometric modality that is being used and the exact application environment in which it will be placed. However, if a variable threshold is decided upon, then different weights of importance will have to be assigned to the numerous criteria that are being utilized.

Finally, the C-level executive should keep in mind that there is a steep trade-off between end user convenience (which means having a very low FAR) and the level of security (which means having a very low FAR) in either both verification or identification applications, with regard to implementing the optimal security threshold setting.

For example, if the convenience of end users is valued over security, this is called a liberal security threshold setting. On the contrary, if security is valued over end user convenience, this is then called a conservative security threshold setting. Most biometric applications lie in between these two extremes.

With regard to the administration subsystem design, this is where the biometrics system administrator has all of their controls located via the control panel. Its design is again heavily dependent upon the type of biometric system that is being used and the particular vendor who designed it.

Finally, in terms of data transmission subsystem design, this is essentially the networking component of the biometric system and will be discussed in greater detail in the next section. Generally speaking, the simpler the data transmission designs, the higher the level of security that can be offered.

Conversely, the more complex a data transmission design, the greater the points of system failure as well as potential security breaches. A very key, but often neglected, component of subsystem design is that of the biometric interchange formats, which spells out how the biometric templates will traverse safely across the network medium in the data transmission subsystem.

Before any biometric system can be fully deployed and declared to be operational by the place of business or organization, the system must first be fully tested to fully ensure that it fully meets the security requirements and, most importantly, that all of the technologies involved are properly functioning in the way that they should be.

Subsystem Implementation and Testing

This is the very next step into the biometrics project management life cycle: that of the biometric system implementation and testing. The testing of an actual biometric system can be broken down into three distinct parts:

1. *Biometric subsystem-level testing*: This is when the biometric devices in the entire system to be deployed are tested first each individually and in a stand-alone mode.
2. *Biometric system integration-level testing*: This occurs when the individual biometric devices in the aggregate system are all joined together into their respective systems and tested as different groups to make sure that all biometric devices within their groups are working harmoniously.
3. *Biometric systemwide-level testing*: This is when the biometric system is tested as an entire whole, or one cohesive security unit, to fully ensure and guarantee 100% that everything is working perfectly.

Throughout the three phases of the biometric system as described above, a key aspect that is of prime importance is that of the quality assurance checks that have to take place. The quality assurance life cycle ensures that all of the bugs and errors found during the system testing have been fully corrected, and the entire biometric system functions fully at the level it was designed and intended to do.

The quality assurance life cycle includes the following:

1. *Verification*: This is the quality assurance process that ensures that the newly implemented biometric system is totally free of errors in the hardware and software.
2. *Validation*: This is the quality assurance check that helps guarantee that the results of the newly implemented biometric system match the original project management plan.

Another type of test that is optional is that of the biometric system prototype testing. This is when a smaller version of the entire biometric system, or its snapshot, is implemented at the place of business or organization. The goal of this kind of test is to see how a small but representative sample of the biometric system will work in a live environment.

The goal of biometric system prototype testing is to fully understand the risks and the type of end user training that will be required before considering a full-fledged biometric system implementation. In other words, this type of testing allows for a business or organization to take certain and calculated *safe* risks to see if a biometric technology will work best to meet their particular security needs and requirements.

Finally, the very last type of test that is required is that of biometric system acceptance testing. This is one of the most important phases of the biometric project management life cycle, as this proves that the actual biometric system is working and fully functional. Under this testing, a representative sample of the end user group is also asked to participate, and if all goes well, the final and sometimes ceremonious signoff occurs, and the biometric system then goes 100% live and operational at the place of business or organization.

System Deployment and Integration

Another key aspect in the biometrics project management life cycle is that of biometric system deployment and integration. With this component, the place of business or organization contemplates whether or not implementing a biometric system as a security add-on to their existing security infrastructure would create a multimodal or a multiple security defense layer.

The key question that needs to be answered here is if the biometric system will be the prime means of security for the business or organization or if the combined security system (which is the existing security system plus the new biometric system) will provide the total security protection.

Very often, however, it is the latter security model that is typically chosen. In this case, it is up to the C-level executives and their respective IT staff to figure out how the new biometric system will integrate with the existing security infrastructure. For this big task to be accomplished, a number of key questions need to be answered:

1. Just exactly at what points will the new biometric system interface with the existing biometric system?
2. Will the newly implemented biometric system take second priority over the existing security infrastructure? In other words, will the new biometric system be considered just as an add-on to the existing security infrastructure?

In this situation, the new biometric system becomes a *technical interface* to the existing security infrastructure system. It can take this shape in three different ways:

1. The biometric system can exist as a tightly woven interface with the existing security system.
2. The biometric system can exist as a loosely woven interface with the existing security system.
3. The biometric system can act as a middleware interface with the existing security system.

With a loosely woven interface, this is merely a simple matchup of the outputs of the existing security system with the inputs of the biometric system. This type of communication is very often used to implement a networking protocol between the two systems.

In terms of a tightly woven interface, the C-level executives and the IT staff, in this case, will decide how the new biometric system will fit into and work seamlessly into the existing security infrastructure. The result is a multimodal security system that operated as one, cohesive unit. However, it should be noted that if the new biometric system will not fit well and easily into the existing security system, modifications will have to be made to the latter in order to accommodate the former.

Finally, with regard to the middleware interface, this is a biometrics management system that is literally sandwiched between the two layers of the existing security system. Middleware can be considered a watered-down version of a full-fledged biometric device, and it has its own functionalities as well.

Middleware

Biometrics middleware can also be considered a cost alternative to a full-blown biometrics system. Biometrics middleware serves the following functions:

1. Acts as an integrating device between various legacy systems
2. Has a free flow of communication with various biometric systems (such as optical-, ultrasound-, silicon-based, etc.)
3. Helps manage various biometric databases
4. Works as a liaison with the networking protocols in the existing security infrastructure
5. Works in an open-source security hardware and software environment

It should be noted that biometrics-based middleware is either client side or server side. For example, with a client-side interface, the biometrics middleware can support a whole host of biometric sensors from all types of vendors.

With a server-side interface, the biometrics middleware assumes all of the backend management of the entire security system, which involves verification and/or identification transaction processing. If a biometrics system is deemed to be very large in nature, then a biometrics middleware known as an *enterprise solutions middleware* is often required.

This type of biometrics middleware solution is either client side or server side or even both at the same time. This type of middleware solution is designed to handle much more complex tasks and functions, such as those of large-scale identification applications.

The Biometric Interface

The interface to the new biometric system and the existing security system can be prone to the security breaches from both outside and inside the place of business or organization, and therefore, the appropriate internal controls need to be applied to mitigate such risks.

The interface can either be a standard out-of-the-box solution or it can be designed entirely from scratch to meet the exact needs of the business or organization. In this case, many customized subcomponents will be required such as a programming language translator (to ensure the communication between the new biometric system and the existing biometric system is that of a common protocol); highly specialized security protocols will be needed as well, such as high-level cryptography for the scrambling and descrambling of the biometric information and data as well as other types and kinds of security data from the existing security system.

The actual biometric system implementation involves the actual installation of the biometric devices and the related networking components and finally training the end users on the proper and accurate use of the biometric system.

An often forgotten obstacle of implementing an actual, live biometric system is logistics, which should be included and required. For instance, dedicated facilities may be required to properly house the biometric system (depending upon the specific nature of the application) and physical changes to the building that houses the business or organization may be required as well.

End User/Administrator Training

Along with end user training, equally important is that of the biometric system administrator training, which consists of the following:

1. An understanding of the specific operation of the entire biometrics system
2. Training in the proper maintenance of the biometric devices
3. Education on how to actually train the end users (this is very often the case if the biometrics vendor is not conducting the actual end user training)

System Maintenance

The very last step in the biometric system project management life cycle is that of the biometric system maintenance. In simple terms, these are the steps that are required to keep the biometric system running in peak order and optimal conditions. Very often, one thinks of cleaning and system tweaking of the biometric hardware in terms of this maintenance component.

However, even collecting the data that are recorded by the biometric system can also be used to keep the biometric system in optimal working order. Such data can be collected from the biometric system logs and transactions reports. For instance, the following types of data are useful in keeping a biometric system in good working order:

1. The statistics related to the total number of users that have been accepted and rejected, the total number of FERs, and also the total number of attempts required by the end user before they are ultimately accepted or rejected by the biometric system
2. The actual number of FMNRs recorded by each biometric device in the entire biometric system
3. The actual FMRs recorded by each biometric device in the entire biometric system

As mentioned previously, a big component of biometric system maintenance is fine tuning, which includes the following duties:

1. Manually adjusting the security threshold levels
2. Implementing additional end user training when needed
3. Calibrating the biometric hardware devices within the entire system
4. Changing or reconfiguring the software that drives the applications of the entire biometric system
5. Tweaking the network performance and bandwidth between the biometric devices and the central server
6. Keeping the mathematical algorithms (especially that of the matching algorithms) totally optimized with the latest upgrades and developments from the biometrics vendor
7. Keeping the security threshold level just at the right point so that the security needs of the business or organization will be perfectly met

Upgrading an Existing System/Fine Tuning

Like software applications, biometrics technology has their fair share of releases and versions. These version number releases can come out for an entire biometric system or for just the individual devices that make it up.

An important consideration to take into account when upgrading to a newer biometric system version is that of template aging. This refers to the degree of dissimilarity between the enrollment template and the end user's actual physiological or biological traits that have changed over time.

This could be caused by the aging process that we all experience or any weight changes or even any injuries. A perfect example of this is that of facial recognition. Over time, a person can go through massive weight loss or massive weight gain, thus rendering the initial enrollment templates as useless.

As a result, the facial recognition system will not recognize the particular individual, thus causing him or her to go through the enrollment process all over again. It should be noted that not all physical and behavioral biometric technologies are prone to template aging.

A good example of this is that of iris and retinal recognition. As reviewed previously in Chapter 2, the biological and physiological structure of the iris and the retina hardly ever changes over the lifetime of the individual. However, if the technology in the biometrics system is advanced enough, for example, if it contains neural network technology, it can create a replication of the original enrollment template using various modeling techniques.

This will prevent the end user from having to re-enroll their templates again after a period. This technique is known as *template/model adaptation*. In the biometric system, it could also be the case that the biometric templates could be set to expire at a certain date, thus rendering the need for re-enrollment by the end users.

Typically, biometric templates are not set to expire, but for some reason or another, the system administrator may establish this particular requirement. The method of expiring biometric templates involves deleting the enrollment templates altogether from the biometrics database and having the end users re-enroll fresh templates again.

System Reports and Logs

Finally, as mentioned in Chapter 2, the reports and logs that can be outputted from a biometric system are an important tool for the system administrator to analyze. These types of reports and logs can consist of the following information and data, depending upon the vendor who manufactured the particular technology:

1. End usernames
2. Details of the verification/identification transactions and their corresponding data
3. The decision (if authorization or nonauthorization was the end result)
4. The security threshold setting that was applied to the decision made in item 3

The following are the benefits of the reports and logs that are outputted by a biometric system:

1. Monitors the performance of the biometric system and the devices that comprise it
2. Keeps track of the version history of the modifications and changes made to the database
3. Creates a repository of all of the data associated with the biometric system (such as performance metrics and statistics) for later analyses
4. Provides a paper-based audit trail of the end users utilizing the biometric system

It should be noted that these reports and logs from the biometric system are designed to assist with further optimization of the biometric system to reach its peak levels and to troubleshoot any type of problems.

System Networking

Another very important component to the overall biometric system architecture and the biometric project management life cycle is that of the networking systems that can be applied. When a C-level executive thinks about a biometric system, the image of a stand-alone fingerprint scanner or even of an optical scanner attached to a laptop or netbook often comes to mind. However, what often gets overlooked, especially by the C-level executives, is that an overall biometric system can be composed of many other different biometric modalities, and all of them can be networked either physically or via a wireless connection with each other.

In fact, many biometric systems today consist of not just one biometric device; they consist of many devices that are interlinked. Because of this interlinking with one another, the biometric devices can literally talk with one another, share templates, and also even transmit other types of biometric information and data.

Network Processing Loads

When the biometric devices are all connected with one another, the transaction processing load for verification and identification applications can be shared with distributed network resources. A perfect example of this is that of an airport application. Today, facial recognition and iris recognition are the two primary biometric modalities that are being used to track and identify the passengers at the major international airports all over the world.

Because of the sheer bulk of the number of passengers and travelers coming in and out of these airports, it is impossible for just one facial recognition device or even just one iris recognition device to do all of the verification and identification. Therefore, many devices are placed at various, strategic locations throughout these airports and are all networked together to ensure that all of the transaction processing occurs among all of the devices.

Also, another key advantage with networking the biometric devices together in such a large environment is that if a potential terror suspect is lurking throughout one of these major international airports and eludes surveillance by some of the biometric devices, the chances are high that he or she will be identified by some of the other biometric devices and be subsequently apprehended.

All these are attributable to the networking of the different biometric modalities and thus possessing the ability to share critical information and data with one another. This airport example is just one type of networking scenario.

A Networking Scenario

The other major type of networking scenario that can take place is when all of the biometric devices can still be networked together, but in turn, all of these biometric modalities in the overall system design are also connected to a central server (and perhaps even more than one server can be utilized, such as the large AFIS fingerprint biometric databases administered by the FBI, where all of the processing of verification and identification applications takes place).

A good example of this is that of physical access entry. Imagine a large-scale business, such as that of a warehouse with thousands of employees. Suppose that fingerprint scanners are installed all over the factory space. Obviously, for one fingerprint scanner to do all of the work of confirming every employee would be excessive (not only that, but it is also very time consuming as well).

However, with the fingerprint devices networked to a central server, all of the transaction processing can take place at this large scale very quickly and easily because the sheer computing power and the resources the server possesses are much greater than just one fingerprint scanner.

Biometric Networking Topologies

Both of these types of networking topologies will be reviewed in detail in this section. As a result, it is very important for the C-level executives and their IT staff to have a broad understanding of the various networking topologies that a biometric system can take up and how that networking will interface with their existing security system.

Thus, the goal of this section is to provide the C-level executive with a basic understanding of the networking concepts and designs that a biometric system can take on and to provide a primer when a biometric vendor discusses about the networking concepts that their particular biometric solution can work with. However, it is important for the C-level executive to keep in mind that the specific design of the networking structure of a proposed biometric system will be heavily dependent upon the new security requirements, as dictated by the needs of the business or organization.

This section will review the following:

1. Data packets
2. Data packet switching
3. Network protocols and TCP/IP
4. Network topologies including peer-to-peer and client–server architectures
5. Routers and routing tables

Data Packets

In the world of networking, and even including biometrics, one of the most fundamental units of connectivity is that of the data packet. For example, every time that an e-mail is sent, or a biometric template is sent to a central server, it is done via the data packet (see the following figure for an example of a data packet).

In other words, for two biometric devices to communicate with the central server or even with each other, the data must first be broken down into smaller bits and travel across the network medium, and upon reaching the central server or the destination biometric device, the data must be reassembled again into a format that the central server or the destination biometric device can decipher and comprehend.

For instance, a regular-size biometric template would be simply too big in terms of size to traverse quickly and easily across the network medium. To accomplish this task, the biometric template must be broken down into very small bits by sending a biometric device, and it is these small bits that are known as the data packets.

These data packets (the broken-up biometric template) travel across the network; once they reach the destination biometric device, the data packets are then reassembled again into their whole form and entirety, so that the destination biometric device can understand the data it has received and take the appropriate actions.

A data packet consists of a header, the payload, and the trailer. The header specifies the destination biometric device (or central server) where the data packet has to go, the payload section consists of the broken-down biometric template, and the trailer possesses the ability via a mathematical algorithm to make sure that the data packet gets to the right destination biometric device or the right central server.

Data Packet Subcomponents

In technical terms, the data packet is broken down into various subcomponents, which are described as follows:

1. *4 bits*: Specifies the version type of the network data packet protocol in which the biometric template resides. Currently, there are two types of network protocol data packets, which are IPv4 and IPv6.
2. *4 bits*: Contains what is known as the Internet header length, and this specifies the exact length of the header of the data packet that contains the biometric template.
3. *8 bits*: Contains the type of service, also known more commonly as the quality of service. This merely refers to what priority level the data packet will have in terms of precedence upon reaching the central server or the destination biometric device. These bits also dictate, to a very small degree, which biometric template has what priority in terms of verification and identification processing.

4. *16 bits*: Contains the length of the data packet by itself. The specific size will vary greatly upon the size of the biometric template.
5. *8 bits*: Specifies the exact time to live, and this is the number of hops that the data packet is allowed to take between servers or biometric devices before it is expired and discarded.
6. *8 bits*: Specifies the network protocol that the data packet consisting of the biometric template will traverse across (most likely it will be the TCP/IP protocol, and this will be reviewed in greater detail).
7. *16 bits*: Possesses the header checksum, and this contains a mathematical algorithm (specifically, the cyclical redundancy checking [CRC] algorithm). This is used to check for the biometric template integrity once it reaches the central server or the destination biometric device. The CRC algorithm also detects any errors that may have occurred during the transmission of the data packet.
8. *32 bits*: Contains the source IP address from which the data packet originates (in this case, it will always be the sending biometric device).
9. *32 bits*: Contains the destination address (in this case, this would be the destination biometric device or the central server).

Data Packet Switching

A very important aspect of biometric network design is what is known as data packet switching. For example, when at least three or more biometric devices are connected together, or when a biometric device is connected to three or more central servers, different network routes may be used to reach the destination biometric device or the central server.

This is done to optimize the flow of data packets in the overall biometric network system design. A key question that often gets asked by C-level executives is, what makes the connection possible among two or more biometric devices or among two or more biometric devices that are connected to a central server?

Network Protocols

In other words, they are all networked together, but what makes it possible for the data packets containing the biometric templates to actually flow freely between the biometric devices and the central servers? The answer lies in what is known as the network protocol. Specifically, a network protocol can be defined as "the format and the order of messages exchanged between two or more communicating entities, as well as the actions taken on the transmission and/or receipt of a message or other event" (*Computer Networking: A Top Down Approach*, Kurose, J. F., and Ross, K. W., 2008, Pearson Education, p. 8).

Today, all biometric devices connected with each other have to use certain network protocols to transmit their data packets back and forth. There are many network protocols existing today. These include very specific protocols for open-source applications, closed-source applications, hardwired connections (this simply means that the two biometric devices are connected by an actual, physical cable), and wireless connections.

TCP/IP

In the world of biometric system architectures, the network protocol known as TCP/IP is what is used most commonly. In fact, this particular network protocol is what is used in most applications that relate to the Internet. TCP/IP is actually a combination of two other network protocols.

TCP stands for transmission control protocol, and IP stands for Internet protocol. This combination protocol of TCP/IP is a very powerful one, and in fact, it goes as far back as the 1970s, when it was first conceived of when the very first prototype Internet came out, known as the APRANET.

TCP/IP consists of four distinct network layers:

1. *Link layer*: This layer consists of the communication tools used in the biometric network so that the data packets that contain the biometric templates can travel from one link to the next link until they reach the destination biometric device or the central server.
2. *Internet protocol*: This is the IP component of the TCP/IP protocol and is primarily responsible for the internal communications between the various network links that make up the entire network segments between the biometric devices or the network segment between the biometric devices and the central server.
3. *Transport layer*: This is the TCP component of the TCP/IP protocol and is primarily responsible for the overall communications between the biometric devices or the biometric devices and the central server.

4. *Application layer*: This is where the overall TCP/IP protocol resides and is responsible for all of the processing services that take place, such as verification/identification transaction processing or template matching processing.

As mentioned earlier at the beginning of this networking section, there are two main types of network designs that are possible. For example, biometric devices can be networked to one another or the biometric devices can be all networked to a central server.

Client–Server Network Topology

Now, we give these two types of network designs specific names. One can be described as a peer-to-peer biometric network, and the other can be referred to as a client–server biometric network. In the client–server biometric network, the server is known as the host, and the biometric devices that connect to it are called the clients (see the following figure for an example of a client–server network topology).

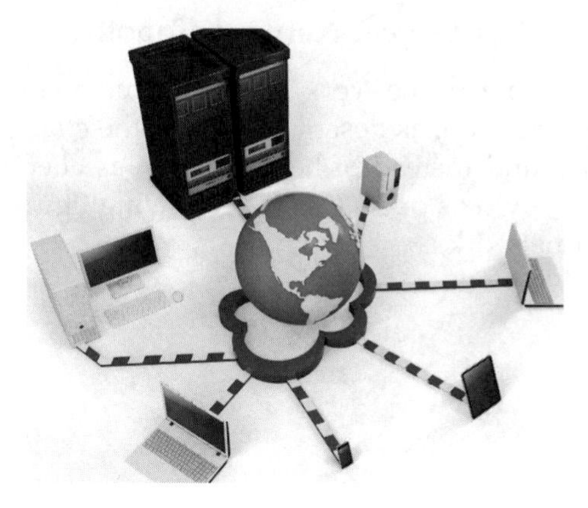

The clients in this type of network architecture merely transmit the biometric templates and other relevant information and data to the host. It is from here that the host then conducts and processes the verification and identification transactions and also contains the databases that house both the enrollment and verification templates.

The host is always on, and the client can either be on or off, depending primarily upon what time during the day they are needed. The transmission of the biometric templates to the host is known as network requests (in other words, the client is asking the host to confirm the identity of the individual in question and to send the result of this transaction back to the client).

Another notable feature of the client–server network biometric topology is that the clients never actually communicate with one another. Rather, they only communicate with the host, which services the network requests. This is so because the host has a unique TCP/IP number attached to it. Based upon this, the clients know which host they need to send their network requests to.

Depending upon the scale and the magnitude of the client–server biometric network topology, often just using one host is not enough to process all of the verification and identification transactions and to store all of the biometric templates in.

Very often, two or more hosts are needed and depending upon the budget of the business or organization and the total number of end users involved, a server farm may be required. In these instances, many hosts (servers) will be required.

Peer-to-Peer Network Topology

In peer-to-peer biometric network, there are no main hosts that are required. In this situation, as described before, the clients are networked with one another, and instead of being known as clients, the biometric devices become known as peers (see the following figure for an example of a peer-to-peer network).

For example, if the processing of verification and identification transactions becomes too much for one peer to handle, then the workload can be transmitted across the other peers, since they are all networked together. There are two key advantages of the peer-to-peer biometric network. They are self-scalability and cost savings.

For example, with the former advantage, each peer adds on processing resource capabilities because the workload can be shared among all of the peers, as opposed to just a few main hosts. Also, since processing of the verification and identification applications is conducted with each peer, the processing time is much quicker as opposed to the client–server biometric network topology. This is because the biometric templates and other relevant information and data have to be transmitted back to the host for the transaction processing to take place.

With regard to the latter advantage, since the peers are also hosts to a certain degree, there is no need to acquire a dedicated, central host. Also, a combination of a peer-to-peer biometric network and a client–server biometric is possible, and such a network is known as a hybrid biometric network.

As discussed previously, packet switching is often used to help the data packets follow the most optimal network route to reach the destination biometric device or a central server in the most efficient and quickest manner possible.

However, packet switching is not used in most typical biometric system network architectures; rather it is only used in the largest of biometric system applications. An example of this is that of the large AFIS databases or the very large international airport environments, or for that matter, if a very large multinational corporation has a massive biometric network system layout, packet switching will most likely be used as well.

Routers

However, to help further optimize the flow of data packets (which contain the biometric templates) in a very large biometric system network topology, routers are often used to help break down the maze of network connections from biometric device to biometric device and from biometric device to central server. In other words, a router can be used to help literally create a group of smaller networks among a huge, overall, biometric system network topology. These smaller networks are known as subnets (an example of a router can be seen in the following figure).

Apart from data packet optimization, routers are used also to help the biometric system administrator to fully map out and understand the logical flow of data packets between biometric devices and central servers. As a result, the flow of data packets will be much more smoother and efficient, and because of this, the transaction processing times of both verification and identification applications will be also just that much quicker, especially in critical, large-scale law enforcement biometric template searches.

Essentially, the primary goal of the router in a large-scale biometric system network topology is to forward the data packets containing the biometric templates along to the right destination biometric device or central server.

This is done by collecting the information and data from the header of the incoming data packets, which are going into the router. Obviously, the router by itself is not going to know the most optimal path to send the data packets along to their ultimate destination, and thus, it uses what is known as a routing table.

In technical terms, the router can be defined as "data table stored in a router that lists the routes to particular network destinations, and in some cases, metrics (the distances) associated with those routes. It is a data file in RAM that is used to store route information about directly connected and remote networks. The routing table contains network/next hop associations" (*Computer Networking: A Top Down Approach*, Kurose, J. F., and Ross, K. W., 2008, Pearson Education, p. 324).

Routing Tables

In simpler terms, the routing table is merely a map inside of the router that tells it where to send the incoming data packets upon which appropriate outgoing network links to ensure that the data packet reaches the right destination biometric device or central server. A routing table can also calculate the total distance that is required to send the data packet to the destination biometric device or central server. To help even further optimize this, the shortest routes are utilized as much as possible (see the following figure for an example of a routing table).

A router consists of the following components:

1. *Input ports*: This is where the incoming data packets coming into the biometric system network topology first enter into, and it is the incoming point that also helps send the data packets into the right direction into the switching fabric.
2. *Switching fabric*: This piece of hardware is located directly inside of the router, and this connects and acts as a bridge between the input ports and the output ports.
3. *Routing processor*: This is similar to the CPU and the RAM of the router; this contains and manages all of the routing tables and helps to a small degree the other various network management tasks.
4. *Output ports*: This port sends the data packets onto its final destination, whether it be the destination biometric device or the central server. Of course, before the data packet is sent out from here, the most optimal path is first calculated by the routing table.

Network Traffic Collisions

Although it does not happen very often, primarily because the biometric network system topology is a highly specialized network that serves only a few main objectives, from time to time, data packets can get backed up at the point of entry into the incoming port. This can literally cause a traffic jam of sorts, thus further exacerbating the verification and identification processing times.

How does such a situation get resolved? There are three primary methods:

1. *Tail drop*: In simple terms, with this methodology, if the router becomes too overloaded with incoming data packets trying to seek entry into the biometric network system topology, the excess of data packets are merely dropped from the incoming queue.
2. *Random early detection*: With this technique, statistical probabilities are calculated and assigned to every incoming data packet. This is used to help determine which data packets should get dropped.
3. *Weighted random early detection*: In this technique, instead of statistical probabilities being calculated, statistical weights are computed and assigned to each incoming data packet to help ascertain which ones should get dropped.

Also, since a router has to examine the legitimacy of the incoming data packets, to ascertain where it needs to send them along in the biometric system network topology, a router can also act as a firewall and help protect the overall biometric system network. This is done when the router discards and permanently deletes any malicious data packets from entering into the incoming port of the router.

REVIEW OF CHAPTER 3

The previous chapter reviewed the key concepts of biometric system architecture, the entire project management lifecycle as it relates to biometric deployment, as well as some of the major networking topologies that can be found in everyday medium- to large-sized biometric applications. However, when one thinks of an actual biometric system, one only thinks of it as a system that provides security to the place of business or organization.

However, a biometric system has its particular *needs* as well, in terms of both maintenance upgrades and fine tuning. In fact, in some ways, a biometric system can be likened to that of the cardiovascular system of the human body.

However, the heart also needs fuel itself, via the coronary arteries, which provide fresh blood to the heart to keep it nourished and healthy. These very same coronary arteries need to be protected and secured so that they do not develop occlusions themselves. If this were to happen, a diminished flow of blood results, thus starving the heart of the nutrients it needs to act as an effective pump.

In this very same regard, the biometric system needs to be secure itself so it is not prone to any type of security breaches. When these breaches do happen, this is also analogous to the occlusions that form in the coronary arteries. The common notion is that since a biometric system itself is a security tool, it does not need any further layers of security in of itself.

However, just as described, this is far from the truth. As mentioned previously, at times, throughout this book, biometrics is just another piece of technology that is also prone to its fair share of hacks and security attacks. However, exactly where are these security vulnerabilities within a biometric system?

Some of the most common attacks can take place in the actual hardware or the software applications themselves. However, normally, these attacks can for the most part be prevented in the future with regular maintenance of the biometric system by installing the required firmware and software upgrades and patches.

Also, hackers have tried to spoof biometric systems by presenting such scenarios as a fake fingerprint or even a fake eye in front of the biometric sensor. However, as discussed earlier in this book, a biometric system only processes physiological and behavioral traits from living tissues.

Another area that is also prone to hacking and security breaches is in the biometrics templates. As reviewed in Chapter 2, biometric templates are just mathematical representations of our own unique physiological and behavioral traits that are used for both verification and identification applications.

If biometric templates are hacked, at least in theory, there is not much that a hacker can do with it—at least not yet. However, there is always a potential that hackers can cause real harm with a hijacked biometric template—especially given the rapid developments of technology that are in their hands. Also, given the societal impacts that biometrics technology possesses, if a biometric database were to be hacked and the templates were indeed stolen, fears among the end users would greatly escalate, and acceptance of the biometric system would greatly deteriorate in turn.

4

An Introduction to Biocryptography

CRYPTOGRAPHY AND BIOCRYPTOGRAPHY

It is very important to further protect the biometric templates—by adding an extra layer of security to them. This solution is primarily answered by the science of cryptography—especially that of biocryptography, which combines both the sciences and technologies of biometrics and cryptography. Therefore, the goal of this chapter is to review the science and principles of cryptography and how it lends itself to the emerging field of biocryptography.

INTRODUCTION TO CRYPTOGRAPHY

Cryptography is a science that dates all the way back to the times of Julius Caesar. In its simplest terms, the science of cryptography is merely the scrambling and descrambling of text or written messages between two individual parties.

These individual parties can also be referred to the sender and the receiver. It is the former that creates the text or the written message that needs to be sent, and in turn, it is the latter that receives the text or the written message and then reads it and appropriately responds.

In normal, everyday communications, we always trust that the individual party who is receiving the text or written message will receive it accordingly, without any type of problem. Although this mostly happens in our daily lives, given especially the high-tech world we live in today, this sometimes does not occur.

When this actually happens, we always assume that the worst has always occurred. However, what is the worse that could happen? The text or the written message could be intercepted by a third party and maliciously used. Again, in normal everyday conversations, while we would normally trust the other party (the receiving party) from keeping the details of the conversation privileged, there is always a chance that a third party could be covertly listening in and use that very privileged information for purposes of personal gain or exploitation, such as that of identity theft.

We can also extend this example to electronic communications of all types. For example, when we hit the *send* button, what assurances do we have that the receiving party will get our message or that it will not be intercepted by a third party? Obviously, we cannot really ensure any type of safety, especially when it comes to electronic communications, like that of e-mail, which gets transmitted all over the worldwide networks and the Internet.

However, the only thing that can be guaranteed is that if any type of message were to be captured by a third party, it would be rendered useless. However, how is this task actually accomplished? It is done by the scrambling and descrambling of the text or the written message. Much more specifically, the text or the written message is scrambled by the sending party, and it remains scrambled while it is in transit, until the receiving party gets the text or the written message.

Message Scrambling and Descrambling

At this point, the text or the written message must be unscrambled for it to make comprehensible sense for the receiving party. For example, a very simple example of this is that of *I LOVE YOU*. The sending party would scramble this message by rearranging the letters as *UYO I VEOL*. This message would then stay in this scrambled format while it is in transit, until it is received by the receiving party.

They would then descramble it, so it would read again *I LOVE YOU*. Thus, if this message were to have been captured by a third party, the content would be rendered useless and totally undecipherable to the third party. This, in very simple terms, is the science of cryptography. It is basically the art of scrambling and, in turn, the descrambling of the text or the written message into a readable and comprehensible format again.

Specifically, cryptography can be defined as "the practice and study of techniques for secure communication in the presence of third parties (called adversaries). More generally, it is about constructing and analyzing that overcome the influence of adversaries and which are related to the various aspects of data confidentiality, data integrity, authentication, and repudiation" (*Computer Networking: A Top Down Approach*, Kurose, J. F., and Ross, K. W., 2008, Pearson Education, p. 683).

Encryption and Decryption

In terms of cryptography, the terms of scrambling and descrambling have much more specific terms associated with them. Respectively, scrambling and descrambling are also known as *encryption* and *decryption*. Thus, for instance, the written message *I LOVE YOU*, when scrambled by the sending party, becomes what is known as the "encrypted message," meaning that the written message has been disguised in such a manner that it would be totally meaningless, or in the terms of cryptography, it would be what is known as *undecipherable*.

Also, encryption can also be further defined and described as "conversion of information from a readable state to apparent nonsense" (*Computer Networking: A Top Down Approach*, Kurose, J. F., and Ross, K. W., 2008, Pearson Education, p. 683). Now, when the receiving party receives this encrypted written message, it must be descrambled into an understandable and comprehensible state of context. This process of descrambling is also known as *decryption*.

So, rather than saying that cryptography is the science of scrambling and descrambling, it can now be referred to as the science of encryption and decryption. There are also specific terms that are used for the encrypted message as well as the decrypted message. For example, the decrypted message, when returned back into its plain or original state of context, which is comprehensible and decipherable, is also known as the *cleartext* or the *plaintext*.

Ciphertexts

When the decrypted message is again encrypted into a state of context, which is totally incomprehensible and undecipherable, this is known as the *ciphertext*. Thus, to illustrate all these, with the previous example, when the sending party creates the written message of *I LOVE YOU*, this is the plaintext or the cleartext.

Once this message is encrypted into the format of *UYO I VEOL*, and while it is in transit, it becomes known as the ciphertext. Then, once the receiving party gets this ciphertext and then decrypts it into a comprehensible and understandable form of *I LOVE YOU*, this message then becomes the plaintext or the cleartext, yet again.

At this point, the question that often gets asked is, how does the sending party actually encrypt the text or the written message, and how does the receiving party then actually decrypt the ciphertext (which is again, the text or written message that is encrypted)?

Well, in its most simplest form, the text or the written message is encrypted via a special mathematical formula. This formula is specifically known as the *encryption algorithm*. Because the ciphertext is now encrypted by this special mathematical algorithm, it would be rendered useless to a third party with malicious intent because of its totally garbled nature.

As the receiving party receives this ciphertext, it remains in its garbled format, until is it is descrambled. To do this, a *key* is used, which is only known by the sending party and the receiving party. In terms of cryptography, this key is also known as the *cipher*, and it is usually a short string of characters that is needed to break the ciphertext.

As it will be examined later in this chapter, interestingly enough, the encryption algorithm is actually publicly known and is available for everyone to use. Therefore, the key, or the ciphertext, must remain a secret between the sending party and the receiving party.

To send the ciphertext between the sending party and the receiving party and to share the keys that are needed to encrypt and decrypt the ciphertext, specific cryptographic systems are needed. Today, there are two such types of cryptographic systems that exist. They are known as symmetric key systems and asymmetric key systems.

Symmetric Key Systems and Asymmetric Key Systems

The primary difference between these two types of cryptographic systems is that the former uses only one key for encryption and decryption, which is known as the private key of the ciphertext. With the latter, two types of keys are utilized for encryption and decryption of the ciphertext, and these are known as the public key and the private key.

We look at both of these cryptographic systems, first starting with symmetric key systems. One of the simplest methodologies in symmetric key systems is that of the Caesar cipher, which can be attributed to Julius Caesar (thus its name).

The Caesar Methodology

With the Caesar methodology, each letter of the text or the written message is substituted with another letter of the alphabet that is sequenced by so many spaces or letters later in the alphabet. To make things simpler to understand, we can denote this specific sequencing as k. So, four letters out into the alphabet would be represented mathematically as $k = 4$.

Let us go back to our example again of *I LOVE YOU*. If $k = 4$, which represents the letter sequencing, the letter that would replace *I* in this plaintext message would be the letter *E*. So, continuing in this fashion, the ciphertext would be translated as *E YOUA EIM*. With the Caesar cipher, a technique known as *wraparound* is possible. Meaning, once the last letter of the ciphertext is reached in the alphabet (which would be the letter *Z*), it wraps around immediately to the start of the alphabet, with the letter *A*.

Thus, if the wraparound technique were needed to finish the encryption of this plaintext message, with $k = 4$, the letter *A* in the plaintext message would become the letter *D* and so on. The value of the k serves as the key in the Caesar cipher, since it specifies how the plaintext should be encrypted over into the ciphertext.

With the Casear cipher, some 25 different combinations or key values can be used. An improvement over the Caesar cipher came with a newer technique known as the *monoalphabetic cipher*. What distinguishes this from the Caesar is that, although one letter of the alphabet can still be replaced with another, there is no exact mathematical sequencing that is required.

Rather, the letters in the plaintext can be substituted at random to create the ciphertext. Again, for example, the plaintext message of *I LOVE YOU* can be written at will and at random as *UYO VOLI E*. With the monoalphabetic cipher, there are more pairings of letters possible. For example, there are 10^{26} possibilities of letter pairings versus only the 25 letter pairings available with the Caesar cipher.

Thus, if a hacker were to attempt a brute force attack on a monoalphabetic cipher (which is just the sheer guessing of the ciphertext for any type of pattern to decipher the plaintext), it would take a much longer time to crack versus the Caesar cipher.

Types of Cryptographic Attacks

However, with both of these types of cryptographic methods just described, there are three types of attacks they are vulnerable to, which are detailed as follows:

1. *Ciphertext-only attack*: With this type of attack, only the cipher text is known to the attacker. However, if this particular individual is well trained in statistics, then he or she can use various statistical techniques to break the ciphertext back into the plaintext.
2. *Known-plaintext attack*: This occurs when the hacker knows some aspect of either the letter pairings; thus, they can consequently crack the ciphertext back into the plaintext.
3. *Chosen-plaintext attack*: With this type of attack, the hacker can intercept the natural plaintext message that is being transmitted across the network medium, and from this, reverse engineer it back into its ciphertext form, in an attempt to figure out the specific encryption scheme.

Polyalphabetic Encryption

Over time, improvements, were made to both the Caesar cipher and the monoalphabetic cipher. The next step up from these two techniques was another technique known as the *polyalphabetic encryption*. With this, multiple types of Caesar ciphers are used, but these ciphers are used in a specific sequence, which repeats again when the overall cipher has reached its logical end the first time in order to finish the completion of the encryption of the plaintext message.

This means that the wraparound technique is also prevalent in this type of scenario. Let us illustrate this example again with *I LOVE YOU*. Building upon the example used previously, suppose that two types of Caesar ciphers are being utilized, such as where $k = 1$, and $k = 2$ (k again denotes the actual Caesar cipher or the sequential spacing of the number of letters later in the alphabet).

The following chart demonstrates this to make it clearer:

Plaintext:	A B C D E F G H I J K L M N O P Q R S T U V W X Y Z
First Caesar cipher, where $k = 1$	B C D E F G H I J K L M N O P Q R S T U V W X Y Z A
Second Caesar cipher, where $k = 2$	C D E G H I J K L M N O P Q R S T U V W X Y Z A B C

The overall cipher algorithm utilized is C1 ($k = 1$), C2 ($k = 2$) where C denotes the Caesar key. Thus, with the example of using *I LOVE YOU*, using the polyalphabetic algorithm, it would be encrypted as *J ORYF BPX*. To understand this further, the first letter in the plaintext is C1, so I is represented as J, the second letter of the plaintext is C2, so C is represented as O, and so on.

The logical end of the cipher algorithm is C2, again, it reaches the logical end of its first iteration, it then wraps around again as C1 ($k = 1$), C2 ($k = 2$) the second time around, then the third time around, until the plaintext message has been fully encrypted.

So, in our illustration of *I LOVE YOU*, there were a total of three iterations of C1 ($k = 1$), C2 ($k = 2$), to fully encrypt the plaintext. Over the years, a more modern form of encryption was developed, known as the *block cipher*. With this method, the plaintext message is put together in one long stream, and from this, it is then broken up into blocks of text of equal value.

Block Ciphers

From here, using a method of transposition, the plaintext message is then encrypted into its scrambled format. Let us illustrate this again with our example used before, but this time, let us assume a block of three characters, mathematically represented as 3 bits, or where $k = 3$.

Plaintext:	I LOVE YOU
Plaintext block:	ILO VEY OUX
Ciphertext block:	OLI YEV XUO
Ciphertext:	OLIYEVXUO

Note that an extra character is added at the end, which is the letter X. This was added so that a complete plaintext block can be formed. As a rule of thumb, if the total number of characters in the plaintext is not divisible by the block size permutation (in this instance, where $k = 3$), then it can be safely assumed that extra characters will be needed to the plaintext for the last block of plaintext to be considered as complete. This is known as *padding*. It should be noted that the most widely used block is where $k = 8$ bits long.

As we can see, even with the simple example from above, block ciphers are a very powerful tool for symmetric key cryptographic systems. After all, it goes through a set number of iterations of scrambling to come up with a rather well-protected ciphertext. However, despite these strong advantages of block ciphers, it does suffer from one kind of inherent weakness.

If it is discovered by a hacker, it can cause rather detrimental damage, with irrevocable results. This vulnerability is that the two blocks of data can contain the exact same data. Let us examine this with our previous example again. As it was illustrated, the ciphertext block was formulated as *OLI YEV XUO*. However, of course, depending upon the actual written context of the plaintext, it is possible that the ciphertext block can contain two or more exact blocks of the same data.

Initialization Vectors

Thus, for example, it would look like this: *OLI OLI YEV*. To alleviate this weakness, systems of initialization vectors (also known as IVs) are used. Although it sounds complex, simply put, it is the technique of creating some further scrambling or randomness within the ciphertext block itself. However, it should be noted that it is not the initialization vector itself that does not promote the ciphertext blocks.

Cipher Block Chaining

Rather, the initialization vectors are part of a much larger process known as *cipher block chaining*, or CBC. Within this methodology, multiple loops of encryption are created to totally further scramble the ciphertext. Here is the how the process works:

1. The initialization vector is created first.
2. Through a mathematical process known as XOR (which stands for eXclusive OR and is used quite frequently to determine if the

bits of two strings of data match or not), the first created initialization vector is XOR'd with the first block of ciphertext data.
3. The first chunk of data that has been XOR'd is further broken down by another layer of encryption.
4. This process is then continued until all of the blocks of ciphertext have been XOR'd and enveloped with another layer of encryption.

Thus, this is how cipher block chaining gets its title. For instance, steps 1–4 as detailed above creates the first loop or chain, the second loop or chain is then next initiated, and so on, until the ciphertext has been fully analyzed and encrypted by this methodology.

Disadvantages of Symmetric Key Cryptography

Now that we have reviewed some of the basic principles of symmetric key cryptography, although it can be a robust system to use, it does suffer from three major vulnerabilities, which are as follows:

1. Key distribution
2. Key storage and recovery
3. Open systems

With regard to the first one, key distribution, symmetric cryptography requires the sharing of secret keys between the two parties (sending and receiving), which requires the implicit trust that this key will not be shared with any other outside third party. The only way that any type of secrecy can be achieved in this regard would be to establish a secure channel.

Although this works very well in theory, in practice, it is not a feasible solution. For instance, the typical place of business or organization would not be able to afford to implement or deploy such a secure channel, except for the very large corporations and government entities. Thus, the only other solution available in this circumstance would be the use of a so-called designated *controller.*

This third party would have to be very much trusted by both the sending and the receiving parties. However, this methodology of trust to create a secure channel can prove to be a very cumbersome task. For example, imagine a place of business. Suppose also that the CEO decides to share the keys of the business with the employees who need access to it at irregular hours.

Rather than trusting the employees explicitly, the CEO could decide to utilize a manager to whom the employees must give the key when they are done with their job duties, and from there, this same manager would then give this key to the next employee who needed access.

Already, one can see that this is a very tedious and time-consuming process, and to compound this problem even more is that the designated controller, in this case the manager, cannot be trusted either because in between the distribution of keys to the employees, this manager could very well give these same keys to a malicious third party. As a result, when this is applied to the world of cryptography, this method does not at all, by any means, guarantee the secrecy of the key that is needed to encrypt and decrypt the plaintext message.

In terms of key storage and capacity, let us take the example of a very large place of business or organization, such as that of a multinational corporation. The problem of using the principles of symmetric cryptography becomes quite simple.

First, since there will be many more lines of communication between the sending and the receiving parties, the need to implement more controllers (and perhaps even more) becomes totally unrealistic as well as infeasible. Thus, the distribution of the keys can become a virtual nightmare.

Second, all of the private keys associated in symmetric cryptography have to be securely stored somewhere, primarily in a database that resides in a central server. As it is well known, primary and central servers are often very much prone to worms, viruses, as well as other types of malicious software. Compounding this problem even more is the fact that if there are many private keys stored onto this central server, the greater are the chances of the central server being hacked into.

A way that these private keys can be stolen is if a piece of malicious code is injected into the intranet of the corporate network, which in turn reaches the database. This malicious code then actually very covertly hijacks and sends back to the hacker these private keys.

Third, when companies and organizations get large, the chances that employees will require remote access to the corporate intranet and network resources become even greater. As a result, the private keys that are used to communicate between the sending and the receiving parties can also be hijacked very quickly and easily by a hacker who has enough experience and knowledge about what they are doing.

Finally, with an open system, private or symmetric cryptography works best only when it is used in a very closed or *sterile* environment, where there are at best only just a few (or even just a handful) of sending and receiving parties. However, this is not the case with *open* or public environments, such as our example of the very large corporation. In these situations, there is simply no way to confirm the authenticity or the integrity of the private keys and their respective ciphertext messages.

Thus, as one can see, private keys and symmetric cryptography simply are inflexible, too costly, and do not scale well for most types of environments. For example, "solutions that are based on private-key cryptography are not sufficient to deal with the problem of secure communications in open systems where parties cannot physically meet, or where parties have transient interactions" (*Computer Networking: A Top Down Approach*, Kurose, J. F., and Ross, K. W., Pearson Education, 2008, p. 687).

Although there will never be a perfect 100% solution that will correct the flaws of symmetric cryptography, there is a partial solution known as *key distribution centers*, which is reviewed next.

The Key Distribution Center

The key distribution center, also referred to as the KDC, is primarily a central server that is dedicated solely to the KDC network configuration. It merely consists of a database of all of the end users at the place of business or organization and their respective passwords as well as other trusted servers and computers along the network.

It should be noted that these passwords are also encrypted. Now, if one end user wishes to communicate with another end user on a different computer system, the sending party enters their password into the KDC, using a specialized software called *Kerberos*. When the password is received by the KDC, the Kerberos then uses a special mathematical algorithm that adds the receiving party's information and converts it over to a cryptographic key.

Once this encrypted key has been established, the KDC then sets up and establishes other keys for the encryption of the communication session between the sending and the receiving parties. These other keys are also referred to as *tickets*. These tickets have a time expiration associated with them, so the ticket will actually expire at a predetermined point in time to prevent unauthorized use, and it would also be rendered useless if it is stolen, hijacked, or intercepted by a third party.

Although the KDC system just described does provide a partial solution to the shortcomings of symmetric key cryptography, the KDC also by nature has some major security flaws:

1. If an attack is successful on the KDC, the entire communications channel within the place of business or organization will completely break down. Also, personnel with access to the KDC can easily decrypt the ciphertext messages between the sending and receiving parties.
2. The KDC process presents a single point of failure for the place of business or organization. If the server containing the KDC crashes, then all types of secure communications becomes impossible to have, at least on a temporary basis. Also, since all of the end users will be hitting the KDC at peak times, the processing demands placed onto the KDC can be very great, thus heightening the chances that very slow communications between the sending and the receiving parties or even a breakdown of the communications system can also happen.

Mathematical Algorithms with Symmetric Cryptography

There are a number of key mathematical algorithms that are associated with symmetric cryptography, and they can be described as follows:

1. *The Needham-Schroder algorithm*: This algorithm was specifically designed for KDC systems to deal with sending and receiving parties from within the place of business or organization, who appear to be offline. For example, if the sending party sends a ciphertext message to the receiving party and after sending the message they go offline, the KDC system could just literally *hang* and maintain an open session indefinitely, until the sending party comes back online again. With this particular algorithm, this problem is averted by immediately terminating the communication session once either party goes offline.
2. *The Digital Encryption Standard (DES) algorithm*: This mathematical algorithm was developed in 1975, and by 1981, it became the de facto algorithm for symmetric cryptography systems. This is a powerful algorithm, as it puts the ciphertext through 16 iterations to ensure full encryption.
3. *The Triple Digit Encryption Standard algorithm (3DES)*: This mathematical algorithm was developed as an upgrade to the previous DES algorithm just described. The primary difference between the two of them is that 3DES puts the ciphertext through three times as many more iterations than the DES algorithm.

4. *The International Data Encryption Algorithm (IDEA)*: This is a newer mathematical algorithm than 3DES and is constantly shifting the letters of the ciphertext message around constantly, until is decrypted by the receiving party. It is three times faster than any of the other DES algorithms just reviewed, and as a result, it does not consume as much processor power as the DES algorithms do.
5. *The Advanced Encryption Standard (AES) algorithm*: This is the latest symmetric cryptography algorithm and was developed in 2000, primarily designed for use by the federal government.

The Hashing Function

Finally, in symmetric cryptography, it should be noted that all of the ciphertext messages come with what is known as a *hash*. It is a one-way mathematical function, meaning, it can be encrypted but it cannot be decrypted. Its primary purpose is not to encrypt the ciphertext; rather, its primary purpose is to prove that the message in the ciphertext has not changed in any way, shape, or form. This is also referred to as *message integrity*.

For example, if the sending party sends its message to the receiving party, the message (or the ciphertext) will have a hash function with it. The receiving party can then run a hash algorithm, and if the ciphertext message has remained intact, then the receiving party can be assured that the message they have received is indeed authentic, and has not been compromised in any way. However, if the hash mathematical values are different, then it is quite possible that the message is not authentic and that it has been compromised.

With the principles of symmetric cryptography, only one key is used to encrypt and decrypt the ciphertext between the sending and the receiving parties. Now, in this section, we look at an entirely different methodology—called asymmetric key cryptography. With this type of methodology, not just one key is used, but rather, two keys are used.

Asymmetric Key Cryptography

These keys are called the public and the private keys and are also used to encrypt and decrypt the ciphertext that is sent between the sending and the receiving parties as they communicate with one another. In the most simplistic terms, asymmetric cryptography can be likened to that of a safety box at a local bank. In this example, normally, there are two set of keys that are used.

One key is the one that the bank gives to you. This can be referred to as the public key because it is used over and over by past renters of this particular safety deposit box, and for other, future renters as well. The second key is the private key that the bank keeps in their possession at all times, and only the bank personnel know where it is kept.

The world of asymmetric cryptography is just like this example, but of course, it is much more complex than this in practice. To start with, typically, in asymmetric cryptography, it is the receiving party that is primarily responsible for generating both the public and the private keys. In this situation, let us refer to the public key as *pk* and the private key as *sk*.

Thus, to represent both of these keys together, it would be mathematically represented as (pk,sk). It is then the sending party that uses the public key (pk) to encrypt the message they wish to send to the receiving party, which then uses the private key (sk), which they have privately and personally formulated to decrypt the encrypted ciphertext from the sending party.

Remember, one of the primary goals of asymmetric cryptography is to avoid the need for both the sending and the receiving parties from having to meet literally face to face to decide on how to protect (or encrypt) their communications with another. Thus, at this point, the question that arises is, how does the sending party know about the public key (pk) generated by the receiving party so that the two can communicate with each other?

Keys and Public Private Keys

There are two distinct ways in which this can be accomplished: (1) the receiving party can deliberately and purposefully notify the sending party of the public key (pk) in a public channel, so that communications can be initiated and then further established; and (2) the sending party and the receiving party do not know anything about each other in advance. In this case, the receiving party makes their public key known on a global basis, so that whomever wishes to communicate with the receiving party can do so, as a result.

Now, this brings up a very important point: The public key is literally *public*, meaning anybody can use it, even all of the hackers in the world, if that is the case. Thus, how does asymmetric cryptography remains secure? It remains solely on the privacy of the private key (sk) that is being utilized. In these cases, it is then up to the receiving party now to share the private key (sk) with any other party, no matter how much they are trusted.

If the privacy of the secret key (sk) is compromised in any way, the security scheme of asymmetric cryptography is then totally compromised. To help ensure that the private keys remain private, asymmetric cryptography uses the power of prime numbers. The basic idea here is to create a very large prime number as a product of two very large prime numbers.

Mathematically put, the basic premise is that it will take a hacker a very long time to figure out the two prime number multiples of a very large product that is several hundred integers long, and thus, gives up in frustration. Even if a hacker were to spend the time to figure out one of these prime numbers, the hacker still has to figure out the other prime number, and the chances that they will figure this out are almost nil.

As a result, only one portion of the (pk, sk) is figured out, and the asymmetric cryptography technique utilized by the sending and the receiving parties still remains intact and secure. In other words, the hacker cannot reverse engineer one key to get to the other key to break the ciphertext. It should also be noted that in asymmetric key cryptography, the same public key can be used by multiple, different sending parties to communicate with the single receiving party, thus forming a one-to-many, or 1:N, mathematical relationship.

The Differences between Asymmetric and Symmetric Cryptography

Now that we have provided a starting point into asymmetric cryptography, it is important at this juncture to review some of the important distinctions and the differences between this and symmetric cryptography. First, with symmetric cryptography, the complete 100% secrecy of the key must be assured, but as has been discussed, asymmetric cryptography requires only half of the secrecy, namely that of the private key (sk).

Although this might seem like just a minor difference, the implications of this have great magnitudes. For example, with symmetric cryptography, both the sender and the receiver need to be able to communicate the secret key generated with each other first, and the only way that this can happen is if both parties met face to face with each other, before the encrypted communication can take place between both parties.

To complicate matters even more, it is imperative that this private or secret key is not shared with anybody else or even intercepted by a third party. However, again, in asymmetric cryptography, the public key can be shared virtually indiscriminately with each other, without the fear of compromising security.

Second, symmetric cryptography utilizes the same secret key for the encryption and decryption of the ciphertext, but with asymmetric cryptography, two different keys (namely the public and the private keys) are both used for the encryption and the decryption of the ciphertext.

In other words, in asymmetric cryptography, the roles of the sender and the receiver are not interchangeable with one another like symmetric cryptography. This means that with asymmetric cryptography, the communication is only one way. Because of this, multiple senders can send their ciphertext to just one receiver, but in symmetric cryptography, only one sending party can communicate with just one receiving party.

Also, asymmetric cryptography possesses two key advantages: (1) it allows for the sending party(s) and the receiving party to communicate with one another, even if their lines of communication are being observed by a third party; and (2) because of the multiple key nature, the receiving party needs to keep only one private key to communicate with the multiple sending parties.

The Disadvantages of Asymmetric Cryptography

However, despite all of this, asymmetric cryptography does possess one very serious disadvantage: compared with symmetric cryptography, it is two to three times more slower than symmetric cryptography. This is primarily so because of the multiple parties that are involved, and the multiple keys that are involved as well.

Thus, this takes enormous processing power and is a serious drain to server power and system resources. Thus far in our review of asymmetric cryptography, we have assumed that the potential hacker is merely just eavesdropping in on the ciphertext communications between the sending and the receiving parties. However, if the potential hacker has a strong criminal intent, they can quite easily listen in on the communications on an active basis and cause great harm in the end.

There are two specific cases in which this can happen. First, if the hacker replaces a public key of his own (mathematically represented as pk') while the ciphertext is in transit between the sending and the receiving parties and the receiving party decrypts that ciphertext with that malicious public key (pk'). However, keep in mind, this scenario assumes that the hacker has some substantial information about the private key (sk).

The second situation arises when the hacker can change the mathematical value of the public key or change it while it is in transmission between the sending and the receiving parties. The exact techniques for protecting the public key in asymmetric cryptography will be discussed.

The Mathematical Algorithms of Asymmetric Cryptography

There are a number of key mathematical algorithms that serve as the crux for asymmetric cryptography, and of course, use widely differing mathematical algorithms than the ones used with symmetric cryptography. The algorithms used in asymmetric cryptography are as follows:

1. RSA algorithm
2. Diffie-Hellman algorithm
3. The elliptical wave theory algorithm

In terms of the RSA algorithm, this is probably the most famous and widely used asymmetric cryptography algorithm. In fact, this very algorithm will serve as the foundation for biocryptography later in this chapter. The RSA algorithm originates from the RSA Data Security Corporation and is named after the inventors who created it, namely Ron Rivest, Adi Shamir, and Leonard Adelman.

As reviewed previously, the RSA algorithm uses the power of prime numbers to create both the public key and the private key. However, using such large keys to encrypt such large amounts of data are totally infeasible, from the standpoint of the processing power and central server resources. Instead, ironically, the encryption is done using symmetric algorithms (such as the ones reviewed previously), then the private key gets further encrypted by the receiving party's public key.

Once the receiving party obtains their ciphertext from the sending party, the private key generated by the symmetric cryptography algorithm is decrypted, and the public key that was generated by asymmetric cryptography can then be subsequently used to decrypt the rest of the ciphertext.

In terms of the Diffie Hellman asymmetric algorithm, it is named after its inventors as well, who are Whit Diffie and Martin Hellman. It is also known as the DH algorithm. However, interestingly enough, this algorithm is not used for the encryption of the ciphertext, rather its main concern is to address the problem of finding a solution for the issue of sending a key over a secure channel.

Here is a summary of how it works, on a very simple level:

1. The receiving party as usual has the public key and the private key that they have generated, but this time, they both are created by the DH algorithm.
2. The sending party receives the public key generated by the receiving party and uses this DH algorithm to generate another set of public keys and private keys, but on a temporary basis.
3. The sending party now takes this newly created temporary private key and the public key sent by the receiving party to generate a random, secret number—this becomes known as the *session key*.
4. The sending party uses this newly established session key to encrypt the ciphertext message and sends this forward to the receiving party, with the public key that they have temporarily generated.
5. When the receiving party finally receives the ciphertext from the sending party, the session key can now be derived mathematically.
6. Once the above step has been completed, the receiving party can now decrypt the rest of the ciphertext.

Finally, with elliptical wave theory, it is a much newer type of asymmetric mathematical algorithm. It can be used to encrypt very large amounts of data, and its main advantage is that it is very quick and does not require a lot of server overhead or processing time. As its name implies, elliptical wave theory first starts with a parabolic curve drawn on a normal x,y coordinate Cartesian plane.

After the first series of x and y coordinates are plotted, various lines are then drawn through the image of the curve, and this process continues until many more curves are created and their corresponding intersecting lines are also created.

Once this process has been completed, the plotted x and y coordinates of each of the intersected lines and parabolic curves are then extracted. Once this extraction has been completed, then all of the hundreds and hundreds of x and y coordinates are then added together to create the public and private keys. However, the trick to decrypting a ciphertext message encrypted by elliptical wave theory is that the receiving party has to know the shape of the original elliptical curve, and all of the x and y coordinates of the lines where they intersect with the various curves and the actual starting point at which the addition of the x and y coordinates first started.

The Public Key Infrastructure

Since the public key has become so important in the encryption and the decryption of the ciphertext messages between the sending and receiving parties and given the nature of its public role in the overall communication process, great pains and extensive research have been taken to create an infrastructure that would make the process of creating and sending of the public keys as well as the private keys much more secure and robust.

In fact, this infrastructure is a very sophisticated form of asymmetric cryptography, and it is known as the *Public Key Infrastructure*, or PKI. The basic premise of PKI is to help create, organize, store, distribute, and maintain the public keys. However, in this infrastructure, both of the private and public keys are referred to as *digital signatures*, and they are not created by the sending and receiving parties; rather they are created by a separate entity known as the *certificate authority*.

This entity is usually an outside third party that hosts the technological infrastructure that is needed to initiate, create, and distribute the digital certificates. In a very macroview, the PKI consists of the following components:

1. *The certificate authority, also known as the CA*: This is the outside third party that issues the digital certificates.
2. *The digital certificate*: As mentioned, this consists of both the private key and the public key, which are issued by the CA. This is also the entity that the end user would go to in case he or she needed to have a digital certificate verified. These digital certificates are typically kept in the local computer of the employee or even the central server at the place of business or organization.
3. *The LDAP or X.500 directories*: These are the databases that collect and distribute the digital certificates from the CA.
4. *The registration authority, also known as the RA*: If the place of business or organization is very large (such as a multinational corporation), this entity then usually handles and processes the requests for the required digital certificates and then transmits those requests to the CA to process and create the required digital certificates.

In terms of the CA, in extremely simple terms, it can be viewed as the main governing body, or even the king of the PKI. To start using the PKI to communicate with others, it is the CA that issues the digital certificates, which consists of both the public and private keys.

The Digital Certificates

Each digital certificate that is generated by the CA consists of the following technical specifications:

1. *The digital certificate version number*: Typically, it is either version number 1, 2, or 3.
2. *The Serial Number*: This is the unique ID number that separates and distinguishes a particular digital certificate from all of the others (this can be likened to each digital certificate having its own Social Security Number).
3. *The signature algorithm identifier*: This contains the information and data about the mathematical algorithm used by the CA to issue the particular digital certificate.
4. *The issuer name*: This is the actual name of the CA, which is issuing the digital certificate to the place of business or organization.
5. *The validity period*: This contains both the activation and the deactivation dates of the digital certificates, in other words, this is the lifetime of the digital certificate as determined by the CA.
6. *The public key*: This is created by the CA.
7. *The subject distinguished name*: This is the name that specifies the digital certificate owner.
8. *The subject alternate name e-mail*: This specifies the digital certificate's owner e-mail address (this is where the actual digital certificates go to).
9. *The subject name URL*: This is the web address of the place of business or organization to whom the digital certificates are issued to.

How PKI Works

This is how the first part of the PKI works:

1. The request for the digital certificate is sent to the appropriate CA.
2. After this request has been processed, the digital certificate is issued to the person who is requesting it.

3. The digital certificate then gets signed by confirming the actual identity of the person who is requesting that particular digital certificate.
4. The digital certificate can now be used to encrypt the plaintext into the ciphertext that is sent from the sending party to the receiving party.

The RA is merely a subset of the CA, or rather, it is not intended to replace or take over the role of the CA; instead, it is designed to help if it becomes overwhelmed with digital certificate request traffic.

However, the RA by itself does not grant any type of digital certificates, nor does it confirm the identity of the person who is requesting the digital certificate. Rather, its role is to help process the requests until the processing queue at the CA becomes much more manageable.

The RA sends all of the digital certificate requests in one big batch, rather than one at a time. This process is known as *chaining certificates*. The RA is typically found in very large, multinational corporations, where each office location would have its own RA, and the CA would reside at the main corporate headquarters.

Finally, all digital certificate requests processed by the RA are also associated with a chain of custody trail, for security auditing purposes. The RA can be viewed as a support vehicle for the CA, in which a mathematical, hierarchal relationship exists.

PKI Policies and Rules

It should be noted that for either the CA or the RA to function properly, it is important to have a distinct set of rules and policies in place. These surround the use of the issuance, storage, and the revocation of the expired digital certificate. Although it is out of the scope of this book to get into the exact details of all these rules and policies, the following is just a sampling of some of the topics that need to be addressed:

1. Where and how the records and the audit logs of the CA are to be kept, stored, and archived
2. The administrative roles for the CA
3. Where and how the public keys and the private keys are to be kept, stored, and also backed up
4. What the length of time is for which the public keys and the private keys will be stored
5. If public or private key recovery will be allowed by the CA

6. The length of the time of the validity period for both the public keys and private keys
7. The technique in which the CA can delegate the responsibilities to the RA
8. If the digital certificates to be issued by the CA are to be used for applications and resources
9. If the digital certificates to be issued by the CA are to be used for the sole purposes of just encryption of the ciphertext
10. If there are any types of applications that should be refused to have digital certificates
11. When a digital certificate is initially authorized by the CA, if there will be a finite period of time when the digital certificate will be subject to revocation

As one can see, based upon the establishment of the many rules and policies that need to be set into place, the actual deployment and establishment of a PKI can become quite complex, depending upon the size and the need of the particular business or organization.

The Lightweight Directory Access Protocol

In terms of the database structure for the digital certificates, this is most useful and effective when the Lightweight Directory Access Protocol (LDAP) servers are utilized. LDAP is simply a database protocol that is used for the updating and searching of the directories that run over the TCP/IP network protocol (this is the network protocol that is primarily used by the PKI).

It is the job of the LDAP server of the PKI to contain such information and data as it relates to the digital certificates, the public and the private key storage locations, as well as the matching public and private key labels.

The CA uses a combination of the end user name and the matching tags to specifically locate the digital certificates on the LDAP server. From that point onward, it is the LDAP server that then checks to see if the requested digital certificate is valid or not, and if it is valid, it then retrieves from its database a digital certificate that can then be sent to the end user. Although all digital certificates that are issued have a finite lifespan when they are first issued, they can also be revoked for any reason at anytime by the PKI administrator.

To accomplish this very specific task, a certificate revocation list, or a CRL, is used. This list is composed of the digital certificate serial numbers that have been assigned by the CA. However, looking at this type of information and data can be very taxing on the system resources and processes.

Therefore, in this regard, it is much easier to reissue the digital certificates as they expire, rather than to revoke them and then having to reissue them again, which would mean that the PKI system administrator would then have to update the CRL.

The Public Cryptography Standards

The PKI is also governed by a body known as the Public Key Cryptography Standards, also known as the PKCS. The first of these standards has been previously described, and they are the RSA encryption standard, the Diffie-Hellman key agreement standard, and the elliptical wave theory. The other sets of standards that define a PKI are as follows:

1. *The Password-Based Cryptography Standard*: This describes how to encrypt a private key with a secret key that is derived from a password.
2. *The Extended Certificate Syntax Standard*: This is merely a set of attributes attached onto a digital certificate which has been assigned by the CA.
3. *The Cryptographic Message Syntax Standard*: This standard specifically outlines how to put the digital signatures into digital certificate envelopes, and from there, put that into another digital envelope.
4. *The Private Key Information Syntax Standard*: This standard directly specifies what kind of information and data should be included into a private key, and how that specific key should be formatted.
5. *The Selected Attribute Types*: This is a detailed list that describes the certain encryption attribute types for the last three standards.
6. *The Certification Request Syntax Standard*: This provides the details for the syntax for the digital certificates. Essentially, this standard simply sets forth the parameters that are needed for the CA to understand the digital certificate request.
7. *The Cryptographic Token Interface Standard*: This is an Application Programming Interface (API) for specifying and handling the cryptographic functions as it relates to the smart cards.

8. *The Personal Information Exchange Standard*: This standard specifies exactly how an end user's private keys should be transported across the network medium.
9. *The Cryptographic Token Information Format Standard*: This standard specifies how the applications at a place of business or organization should interface with smart cards.

In the world of PKI, it should be remembered that the public keys and the private keys (also known as the digital certificates) are created instantaneously and all of the time. In fact, public keys and private keys are everywhere in a PKI, even when one establishes a Secure Shell (SSH) connection over the Internet with their particular brand of web browser (this typically uses 128-bit encryption).

In fact, there are even public keys and private keys in the PKI that are only used once, terminated, and are literally discarded away. These types of public keys and private keys are known more commonly as *session keys*. Public keys and private keys are nothing more than computer files.

Parameters of Public Keys and Private Keys

However, before the actual public key or the private key can go out, it needs to have certain parameters that are specified to before it can be created, and these are as follows:

1. The type of mathematical algorithm that should be used (as described previously)
2. How many bits of data the public keys and the private keys should be composed of
3. The expiration date of both the public keys and the private keys

To keep the hackers at bay, it is equally important that not all of the public keys and the private keys are used all the time in the communication process between the sending and the receiving parties. It is also important to keep the public keys and the private keys fresh, or in other words, it is important to introduce randomness into the PKI.

Such randomness is known as *entropy*, and this entropy is created by what is known as *random number generators* and *pseudo-random number generators*. Also, in a PKI, there are different classes of both public keys and private keys. Here is a list of just some of these classes of keys:

1. *Signing keys*: These are the keys to create the digital signatures.
2. *Authentication keys*: These are the keys that are created to authenticate computers, servers, and the receiving parties and the sending parties with one another.
3. *Data encryption keys*: These are the keys that are used to encrypt the files.
4. *Session keys*: These are the types of keys that are used to help secure a channel across an entire network for only a very short period.
5. *Key encryption keys*: These types of keys literally wrap the ciphertext to provide further protection between the sending and the receiving parties.
6. *Reof key*: This is the master that is used for signing all of the other public keys and private keys that originate specifically from the CA.

How Many Servers?

In the PKI, especially in small- to medium-sized businesses, very often, only one server is utilized to distribute both the public keys and the private keys to the employees within the place of business or organization.

However, the primary disadvantage with this is that these types of key servers can literally become the single point of failure, if the server breaks down, or worse yet, it is even hacked into.

To help alleviate this problem, these types of businesses can have multiple, redundant servers, but this too can become a huge expense, especially for the small- to medium-sized businesses. Thus, as a result, the best option for the small- to medium-sized businesses as well as even the large multinational corporations is to outsource the entire PKI to a third party, such as Verisign.

This approach can be considered as a hosted one and can greatly save on information technology (IT)–related expenses, and security is at the forefront of these hosting parties. Another key issue that is related to public key and private key distribution in a PKI is that of the setting the security policies for both the public keys and the private keys and which mechanisms are used for their storage as well as even furthering the securing of the public keys and the private keys themselves (this is a topic for what is known *virtual private networks*, or VPNs.

Security Policies

The exact mechanisms as to how to exactly establish a specific security policy is not the realm of this book, but the security policy should cover at minimum the following key issues as it relates to public key and private key generation and distribution:

1. Who are the individuals authorized to access the key server (assuming that the place of business or organization has one)?
2. Who within the place of business or organization is allowed to use the public keys and the private keys?
3. What types of ciphertexts and content messages and even corporate data can use the encryption methods that are provided by the public and the private keys?
4. If the public key and private key generation and distribution processes are outsourced to a third party, who at the place of business or organization has the authority to issue public keys and private keys after they have been generated?
5. What are the specific requirements for the employees within the place of business or organization to obtain both public keys and private keys?

Securing the Public Keys and the Private Keys

Related to the issue of both public key and private key security is that of how to secure the keys themselves. Broadly speaking, there are two ways on how this can be handled:

1. Key escrow
2. Key recovery

The former refers to the storage of the public keys and the private keys at a safe location, and the latter refers to breaking up the public keys and the private keys at the point of origin (which would be when the ciphertext is sent from the sending party), and putting them back together again at the point of destination (which would be when the receiving party receives the ciphertext).

Message Digests and Hashes

However, in today's business world, both of these methods are infeasible due to security concerns. Today, the preferred method by most CIOs is using message digests and hashes. Both of these refer to the same concept and are often used synonymously with each other. Essentially, it is a fixed length of literally data nonsense, such as

RTYHDHDHjjjdd8585858hd0909344jdjdjdjMNGDfsweqwecbthrdn*&^%gh$

This is to help ensure the integrity of the ciphertext while it is in transit across the network medium. In other words, this is proof positive for the receiving party who receives the ciphertext from the sending party that the ciphertext has remained intact while it has been in transit, and it has not been altered in any way, shape, or form.

The hash or the message digest can be viewed as a fingerprint of the ciphertext. The hash is actually created at the point of the receiving party, and it is then calculated again via a mathematical algorithm utilized by the receiving party.

If the mathematical algorithm used by the receiving party generates the same type of garbled data message such as the one shown in the example, then the receiving party can be 100% sure that the ciphertext they have received from the sending party is the original ciphertext at the point of origination and has remained intact.

Security Vulnerabilities of Hashes

However, a major security vulnerability of using hashes is that they can even be altered while they are en route. In other words, a hacker can intercept the ciphertext and its associated hash, alter both, and create a brand new ciphertext and a brand new hash. As a result, the receiving party is then fooled into believing that this new, altered ciphertext and the new altered hash are the originals sent by the sending party, while the hacker keeps the actual ciphertext and hash that were generated the first time around.

To fix this major security vulnerability, the ciphertext is combined with a *secret key* at the point of origination first, then the hash is created. As a result, this hash will then contain specific information and data about the secret itself. As a result, the receiving party can be even further ensured that the ciphertext they have received is the 100% original sent by the sending party.

This is so because even if the ciphertext, the hash, and the associated secret key were to be intercepted, there is very little that a hacker can do to alter the ciphertext and its associated hash because they have to have the information and data about the secret key, which is something they will never gain access to.

Virtual Private Networks

VPNs can be used as a much more sophisticated form of protecting the integrity of the ciphertext when it is sent from the sending party to the receiving party, and vice versa. A VPN is essentially a dedicated network in and of itself. It is highly specialized, with the main intention of highly securing the flow of communication between the sending and the receiving parties (an example of a VPN can be seen in the following figure).

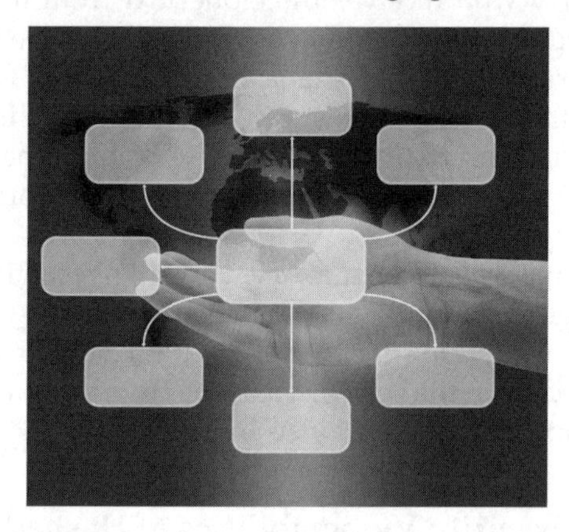

In its most simple form, the VPN takes the data packet in which the ciphertext resides (the ciphertext cannot just be sent by itself—it needs to have a vehicle to travel across the network medium, which is namely the data packet), and then further encrypts the data packet as needed.

This encryption is actually just another data packet. Meaning, the data packet that contains the ciphertext is further wrapped inside by another data packet, providing that extra layer of protection. This process is technically known as *encapsulation*.

Once this process has been accomplished, the VPN then establishes a dedicated network connection, or even a dedicated network channel, in which the encapsulated data packet can be sent. Although this specific connection uses public network infrastructure and related systems, this special type of network connection established and used by the VPN cannot be seen by others, as this connection cannot be picked up by network sniffers.

IP Tunneling

The establishment of this special type of network connection by the VPN is also known as *IP tunneling*. The following example best demonstrates how the scheme of IP tunneling actually works. First, assume that there is an employee with a laptop who wishes to connect to an internal server that is located deep within the company network.

Further assume that this laptop the employee possesses has a TCP/IP address of 1.2.3.4 and that the corporate server the employee wishes to connect to has a TCP/IP address of 192.168.1.1 and that it cannot be reached by any other public access mechanism.

For the individual's laptop to actually reach the internal corporate server, it must first penetrate a VPN firewall/server that has a public address of 10.11.12.13 as well as possessing a private TCP/IP address of 192.168.1.2. To fully establish the connection between the individual's laptop and the internal corporate server, a VPN tunnel must first be established.

Here are the steps to accomplish this:

1. The VPN client will connect to the VPN server/firewall through a specific network interface.
2. The VPN firewall/server will allocate a brand new TCP/IP address (such as 192.170.2.60), from the VPN's server/firewall subnet. This particular technique creates the virtual tunnel through which the encrypted data packet can reach the end of the other VPN tunnel endpoint.
3. Now, when the individual's laptop (also known as the VPN client) wishes to communicate with the corporate server, it prepares its data packets to be sent to the TCP/IP address of 192.168.1.1, the data packets are then encrypted and then further encapsulated with an outer VPN packet (such as IpSEC packet), and relays these further, enhanced data packets to the VPN firewall/server with the public TCP/IP address of 10.11.12.13 over the public Internet. This extra layer of protection is so well encrypted that even if a hacker were to intercept these particular data packets, there is nothing that they can do to reverse engineer the data packets to garner information and data that reside in them.
4. In this stage, when the encapsulated data packets reach the VPN firewall/server, it then extracts the inner data packets, decrypts them, and then further notes the destination TCP/IP address is 192.168.1.1, which is the actual corporate server, and from this point onward, forwards these particular data packets to the corporate server.

This is the process of how this particular employee's laptop can be reached, connected, have access, and communicate with the corporate server that resides deep in the heart of the corporate network. As one can see, the beauty of the VPN is that to an outsider, particularly a hacker, it looks and appears as if the communications of the data packets (such as in this example, between the employee's laptop and the corporate server) is transpiring across the same network segment, when in reality, the actual and real communications takes place across hidden, secret, and covert communications paths.

Mobile VPNs

As the explosion of smartphones takes place in today's world, and especially in today's corporate setting, the use of what is known as mobile VPNs will also soon proliferate at an equally greater pace.

The traditional VPN model is a stationary one, meaning it can only be used at one type of geographic location, such as the one depicted in the example above. However, in today's wireless world, stationary VPNs and the technology that is associated with them simply will not be able to keep up with the quick pace of mobile technology.

Another perfect example of this is that of mobile devices used by law enforcement agencies. Today, law enforcement officers are equipped with various types of mobile devices that allow them to confirm the identity of the suspect whom they have apprehended. However, rather than using the traditional ink-and-paper approach, biometrics are being used to provide a quick determination and turnaround time for the identification of the apprehended suspect.

In these particular mobile applications, the endpoint of the VPN is not fixed upon and located at a single TCP/IP address; instead, the VPN connections roam across the various networks such as data networks from cellular carriers or between multiple WiFi access points.

In these cases, if traditional VPN models were to be utilized, the virtual private connections would continuously get broken at each different location.

From this point onward, the mobile VPN software handles the necessary network authentication and maintains the network tunneling sessions in a way that is totally transparent to the mobile application and the end user.

Is the VPN Worth It?

To the very large multinational corporations, and especially to the CFOs and the CIOs at these places of businesses and organizations, the ability to afford a high-end VPN solution comes to them at virtually no cost, not literally speaking, however, as there is a true cost that is incurred by these businesses and organizations. But for these types of companies, purchasing such a high-end system is just merely a capital expenditure that can be easily written off as a mere deduction when it comes to report taxes in the next fiscal year.

However, for the small- to medium-sized businesses, obtaining a high-end VPN solution can be a very expensive proposition to undertake. For the CFOs and the CIOs of these much smaller businesses and organizations, obtaining a high-level VPN solution can be an extraordinary drain on the IT budget.

However, as the world is now turning more toward the use of biometrics-based VPN solutions, these small- to medium-sized businesses and organizations, at some point, will have to acquire a VPN solution just to keep at a minimum the cat-and-mouse game among the end users, the vendors, and the hackers.

Conducting a VPN Cost–Benefit Analysis

What do the CIOs and the CFOs have in their arsenal to help them to decide which VPN solution is affordable but will also be the best for their place of business or organization? One such technique is a very commonly used financial analysis methodology called a *cost–benefit analysis,* or CBA.

As it relates to the VPN, as the CFO or CIO for your business or organization, you first need to determine if you even have the funds needed to acquire a biometrics-based virtual private solution. The key financial questions to be asked and analyzed include the following items:

1. Determine your actual operating costs to run your current IT infrastructure.
2. Calculate the real and the tangible benefits that the biometrics-based VPN solution will bring to your business or organization as opposed to your existing IT security infrastructure.
3. Calculate the capital requirements as well as the operational costs of the new biometrics-based VPN solution.
4. Spread the financial costs of the estimated life of the new biometric-based VPN solution, but it is very important to assign the capital expenditure costs to the specific years.
5. Ascertain and determine the cost differences between the operating costs of the existing security system in place and the brand new biometrics-based VPN solution.
6. Based upon the financial numbers calculated above, you can now easily calculate the return on investment, or ROI, as well as the

internal rate of return, or IRR, of the biometrics-based VPN solution as well as the net present value. If both of the key financial statistics reveal a positive number, then the biometrics-based VPN is definitely worth considering to acquire.

Now, with the above key financial questions solidly answered, as the CFO and the CIO, you can now conduct the true cost–benefit analysis, and this should include the following components:

1. All of the costs that are associated to keep the biometrics-based VPN running and the associated upkeep costs
2. All of the capital expenditures related to the lifetime of the biometrics-based VPN project
3. The actual costs of hiring and keeping the employees to maintain the VPN system
4. The calculations of the new revenues that the biometrics-based VPN system will bring to the place of business or organization
5. Any other hidden costs it will take to implement the new VPN system
6. Any other IT budget cost savings that can be garnered by the new biometrics-based VPN system

With all of this in mind, as the CIO and the CFO, it is also important to stay away from what is known as the *scope–creep* phenomenon. In other words, just implement the biometrics-based VPN system for the purposes it was intended to do, and not any more.

Otherwise, your operating costs will get greatly out of hand. The rules of IT project management must be strictly enforced. As the CIO, once you have discovered through the processes detailed above that a biometrics-based VPN solution is best for your place of business or organization, you must follow a very orderly approach in deploying the VPN system.

Implementing a VPN

Simply throwing the system at the IT team and having them figure out how to deploy the actual system at your place of business or organization will simply not work. Rather, a very strong planned and iterative approach needs to take place. Such a controlled process involves the following steps:

1. Analyzing
2. Pilot testing

3. The limiting of the full-blown biometrics-based VPN only in stages
4. Conducting a complete VPN rollout once the stages described above have been completed and everything is set in place

It should be kept in mind that part of this iterative process includes conducting a very detailed systems and requirements analysis, which includes carefully reviewing and examining the following major requirements:

1. VPN security requirements
2. VPN application program requirements
3. VPN user access requirements
4. VPN requirements
5. VPN performance requirements

In terms of the security requirements for your new biometrics-based VPN system, as the CIO, you will need to have an extremely firm grasp of the following:

1. An understanding of the types of information system requirements
2. A classification of the IT assets to actually determine what needs to be protected

In terms of the information assets classification for the biometrics-based VPN, the key item to be determined is that of the tangible value of the assets to three very important entities:

1. Your place of business or organization
2. Your business associates (such as your suppliers)
3. Most importantly, your customers

To fully understand and quantify the value of your IT assets, as the CIO, there are three types of risks that need to be ascertained as well:

1. Integrity risk: The data packets that get sent through the VPN system have qualities of both a commodity and a weapon, such as
   a. Data loss
   b. Data alteration
   c. Data theft
2. Confidentiality risk: The data packets that get transmitted back and forth between the VPN solution can contain very private and confidential data about the place of business or organization and are very prone to being lost, stolen, or hacked into and used for malicious purposes by a hacker

3. Availability risks: This is a grave risk and can occur when the biometrics-based VPN totally becomes unavailable to the place of business or organization, and this is known as a denial of service attack, when a hacker totally floods the VPN system and overloads it.

The Components of a VPN Security Policy

In implementing a biometrics-based VPN system, formulating and developing a very sound and "airtight" security policy are of prime importance. The components of a very strong VPN-based security policy should include some of the following components:

1. Access rights:
    a. Which employees should have access to what types of resources
    b. When, where, and how often is access allowed through resources via the biometrics-based VPN
2. Access control rights, which include the following:
    a. IP address source
    b. Data packet content and destination from within the biometrics-based VPN
3. VPN management responsibilities, which include the following:
    a. Who will administer and oversee the biometrics-based VPN solution
    b. Who will enforce the security of the biometrics-based VPN
    c. Who will authorize the issuance and the distribution of the digital certificates
    d. Who will perform the registration activities
4. The types and the degrees of the types as well as the degree of encryption which is required:
    a. Deciding upon the types of and kinds of IPSec Protocol settings and options that are required
    b. The management and distribution of the public keys and the private keys
    c. The length of time for digital certificate activity and expiration
5. VPN endpoints: This simply involves where the IP tunneling will be routed through:
    a. Gateway to gateway
    b. Gateway to desktop
    c. Desktop to desktop

In terms of a biometrics-based VPN system, all types of programs and applications that run in your place of business or organization need to be double checked to make sure that they will be compatible to the new VPN hardware and software. Such programs and applications need to be closely examined, which include:

1. Database access and maintenance programs
2. Mainframe access through terminal emulators
3. Any type of software development tools and their respective databases
4. All sorts of dynamic web content generators used for intranet development
5. Any type of document sharing program
6. All remote server administration hardware/software
7. All backup as well as remote backup tools utilized at your place of business or organization

End Users and Employees

With regard to the biometrics-based VPN system, the concern of users and employees should be of prime concern as well. As the CIO of your business or organization, the following end user requirements need to be taken into consideration before the biometrics-based VPN system is deployed:

1. The total number of remote office locations your place of business or organization is intending to fully support in a biometrics-based VPN system
2. Where the existing corporate networks, such as the local area networks and the wide area networks, cross and intersect with one another
3. How many end users will need to be supported by the biometrics-based VPN system, including regular employees, business partners, and contracted employees
4. The type of end users the biometrics-based VPN must support, such as the following:
   a. Software developers
   b. Sales force employees
   c. Facilities personnel
   d. Office personnel
   e. Customers
   f. Contractors

g. Business partners
h. Suppliers/vendors

As the CIO, it is equally important to consider the type of access to the applications all of your end users will be needing, such as

1. Access to particular network directories
2. Access to all types of corporate and remote servers
3. Access to all types of network resources
4. Mainframe access
5. Database access to the software
6. Web-based and HTTP support
7. E-mail server support
8. Client–server applications
9. Mainframe terminal emulation

The Network Requirements

Above all, to the CIO, determining and ascertaining the biometrics-based VPN system network requirements is of prime importance above anything else and include the following issues that need to be taken into consideration:

1. The design and topology of the network layout and structure
2. The various access points into the overall corporate network and the intranet
3. Any type of dynamic protocol support
4. Internal protocol service requirements
5. All and any existing routers, firewalls, and proxy servers that exist in both inside and outside of the place of business or organization
6. All types of any existing authentication rules that exist in the place of business or organization
7. Any and all business applications that will cross into the boundaries of biometrics-based VPNs
8. All bandwidth requirements set forth by the place of business or organization
9. Any cryptographic processing requirements
10. IT staff and support requirements
11. Full scalability of critical IT devices such as
    a. Authentication servers
    b. Database servers

c. Web servers
d. Any and all mapped driver and directories
e. Corporate e-mail servers and gateways
f. Corporate and end user FTP servers
g. Remote network/server administration

Finally, as the CIO, evaluating the performance of the biometrics-based VPN is equally, if not more, important, and these items need to be reviewed such as

1. The cryptographic hardware accelerator support
2. The clustering of servers for scalability
3. All types of quality of assurance service levels for all of the corporate servers, extranets, and intranets
4. The importance of backup and redundant devices to ensure full and complete nonstop processing of the information and data for the place of business or organization
5. Any expected growth in the network bandwidth requirements

As the CIO of your place of business or organization, more than likely, you will be purchasing a biometrics-based VPN system solution that comes out of the box and is ready to install, perhaps even with the help of an outside consultant.

Building Your Own VPN

If you are one of those more daring types of CIOs who likes to design and implement your own biometrics-based VPN solution, there are a number of key things that you need to be aware of. Although it is out of the realm of this chapter to go over completely the specific needs that must be considered when making your own biometrics-based VPN system, a few critical points are offered here.

First, as the CIO, it is very important to keep in mind the three levels of IT systems that are greatly impacted by the addition of a biometrics-based VPN system. These three items are typically the web servers, the application servers, and the database servers.

However, this is also known as a three-tier advantage and possesses the following strengths:

1. IT is greatly enhanced.
2. The server scalability is much easier to implement.

3. The centralization concepts allows for the IT staff to control and to secure programs and servers.
4. Well-defined software layers at the place of business or organization's software model allows for a quick responsiveness from the IT Team from an ever changing business environment.
5. Any type of existing mainframe services can be reused throughout over time.
6. Biometric-based VPN systems based on an open-source software model allows for the places of business and organizations to incorporate new technologies very quickly and readily.

Impacts to the Web Server

If you are going to design your own biometrics-based VPN solution, the first line of impact will be that of the web server, which can be seen in the following figure. Although many CIOs have heard of the term *web server* before, there is a special type of web server that merely responds to the hypertext transport protocol (HTTP) requests for the HTML pages it delivers to your customer's web browsers.

A key example of this would be your company catalogue that your customers are trying to view online. To improve the response time of the web servers and to deliver high-end quality web pages to your customers, there are three types of processing that can be utilized through a biometrics-based VPN system:

1. *Symmetric multiprocessing, or SMP*: This is when a web server relies upon a single multiprocessor as well as a single memory capability to accomplish the workload that has been assigned to them, but the scalability of the SMP is a very serious limitation.
2. *Massively parallel processing, or MPP*: This is when the role of the logic chip has its very own processor as well as its own memory

subsystem to process information and data. These are all threaded to the nodes and collect the results on the other end. As one can tell, these types of processors are very expensive but also offer limited scalability to the biometrics-based VPN system.

3. *PC clustering*: In a biometrics-based VPN system, this type of approach bridges the first two processing techniques as described above. This type of system can contain unlimited subprocessors, all connected via a fiberlink, or what is known as an optical network. With this type of system, each SMP runs its own version of the operating system and can provide strong backup features. However, the biggest drawback to this type of system is that it is quite cost prohibitive.

Impacts to the Application Server

The second level of the IT system that will be greatly impacted by the adaptation and the implementation of a biometrics-based VPN system is that of the application server, which can be seen in the following figure. Simply put, these types of servers store, manage, operate, and control the specialized software components that are relevant to your business or organization.

When implementing a biometrics-based VPN system, as the CIO, you may also want to consider implementing a trusted host. This simply makes use of access control lists (or ACLs) in an effort to thwart hacking attempts at running or installing various application and software programs, provided that they do not have the authority to be installed onto the application server.

It should be noted that access control information that is stored upon the application servers should be stored on the most garbled form possible and never stored as cleartext anywhere. As the CIO, you should remember not to place too much work on the type of application server, as in turn, this could place a great strain on the biometrics-based VPN system by adversely impacting the inflow of data packets.

Impacts to the Database Server

Finally, the third level of IT system that could be impacted by the implementation of the biometrics-based VPN system is that of the database server, which can be seen in the following figure. For your business or organization, these store the product, purchasing, and customer data as well as all of the other data warehousing techniques that are contained within it.

For the database servers, the object-oriented database management system, or the traditional relational databases could be used. However, regardless of this choice, these specific types of database software must be *prepared* for the implementation of a biometrics-based VPN system. When the biometrics-based VPN is finally installed at your business or organization, it is absolutely imperative to make sure that all company information and data are encrypted across all levels, which includes the field level, the row level, the table level, as well as the entire database level.

Impacts to the Firewall

Finally, as the CIO of your place of business or organization, another key component that cannot be overlooked is that of the firewall—in fact, many security experts agree that putting the biometrics-based VPN in front of the firewall (shown in the following figure) is one of the best choices that can be made. For example,

1. A single set of rules allows you, as the CIO, to control the flow of data packets in and out of the biometrics-based VPN
2. A single point of administration helps you, the CIO, to manage changes quickly as your business or organization dictates
3. Finally, by having a single point of access at the firewall level, this allows you, the CIO, to secure and control your IT systems very responsibly as well as responsively

Now that we have looked upon some of the critical issues that are important to you, the CIO, in terms of some of the major design considerations when creating your own biometrics-based VPN solution, we now need to look at another key aspect of security—testing of your biometrics-based VPN system.

VPN Testing

It does not matter if you have bought it off the shelf or had your IT team design the whole thing from scratch. Once your biometrics-based VPN system is in place, as the CIO, you need to security test the whole system from a number of different angles, which includes providing different security alternatives such as the following:

1. Attacks against biometrics-based VPN solutions include
   a. Always destroy the plaintext.
   b. If you use what is known as virtual memory, make sure that the plaintext that is stored in them always gets destroyed.
   c. From the biometrics-based VPN system, delete any end users passwords from memory after the dialog box closes.
   d. Make sure that the encrypted ciphertext are both protected with strong keys, not just a strong one and a weak one.
   e. Do not let the master keys be governed by the mercy of your session keys.
   f. As much as possible, make sure that every possible method in which a hacker can guess or learn both your public keys and private keys can be mitigated.
2. Attacks against biometrics-based VPNs hardware through the following methods:
   a. Measuring power consumption usage
   b. The ability to measure radiation emissions from *other side channels* in the biometrics-based VPN
   c. Fault tolerance analysis
   d. Purposely and deliberately introducing faults to determine both the public keys and the private keys
3. Attacks against trust modules: As much as you have to be aware of the technological threats to your biometrics-based VPN system, as the CIO, you have to be aware of the human threats that could take place, with the end result of the biometrics-based VPN system being totally compromised. Such human threats include the following:
   a. Which personnel in the place of business or organization can be trusted?
   b. How does the biometrics-based VPN confer trust among the IT staff?
   c. How far should this trust be extended?

   d. How do you, as the CIO, avoid the possibility of collusion between the end users of the biometrics-based VPN system?
   e. As the CIO, what checking mechanisms can you put into place to make sure that human collusion safeguards are indeed being enforced?
   f. How do you avoid the human temptation that you can fully 100% trust all of your IT staff at your place of business or organization?
   g. Making sure that the version control to the software code changes made to the software applications of the biometrics-based VPN system is firmly locked down and secure.
   h. Getting rid of the notion of convenience versus security for the IT staff as well as the entire business or organization.
   i. Making sure that your biometrics-based VPN system is as much 100% irreversible engineered as possible.
   j. Having the false sense of security that a solid trust model is always going to be 100% fail safe and will never be needing any improvements whatsoever. Truth be told, improvements will always have to be made, and the security trust model implemented will always have to be updated to keep up with the so-called cat-and-mouse game against the hackers (this simply means that being at least two to three steps ahead of the hackers as they devise new hacking techniques in just a matter of hours or even minutes).
4. Attacks against failure recovery: Failure recovery describes the system's capability to have the ability to maintain high levels of security in the face of any component breakdown or failure. In other words, to the CIO, this becomes the classic war of functionality versus security—in other words, that fine line has to be drawn and balanced of what is more important to your place of business or organization in the face of catastrophic breakdown: your customer's convenience or the businesses backup and recovery plans?
5. Attacks against cryptography: Despite having the best possible biometrics-based VPN system in place, and the best cryptographic tools to go with it, cryptographic flaws can and still do happen. For instance, to help protect against these, some of the following steps should be taken:
   a. Keep the mathematical values as random as possible, all the time.

b. Make sure that the cryptographic digital signatures are strictly correct about the parameters they need and should be in.
c. From the standpoint of the CIO, you should know and understand the appropriate and the inappropriate uses of the biometrics-based VPN protocols.
d. Establishing the biometrics-based VPN testbed: As a CIO, you must remember that your biometrics-based VPN must be able to communicate and process the data packets with any other biometrics-based VPN systems not only at your place of business or organization, but at other business establishments that may also possess such a VPN. To allow for this free flow of communications between biometrics-based VPN systems, the communications become the corporate intranet and the Internet becomes of paramount importance. To accomplish this task,
    i. Make sure that you dedicate PC's for each individual client.
    ii. Issue test versions of the X.509 digital certificate for each end user and each network resource requiring access to the biometrics-based VPN system.
    iii. Use trusted Internet Service Providers (ISPs) to allow for access from the remote locations, especially where remote connectivity is required.
    iv. If possible, exchange a few of the digital certificates to the test population group at your place of business or organization.
    v. Never use your own corporate intranet for the testing of the biometrics-based VPN system between the client PCs.
    vi. Seriously and strongly consider putting test machines to test your biometrics-based VPN system at different offsite locations.
    vii. If possible, as the CIO, try to use some test cases for both the network gateways and the clients. This will enable you and your IT staff to compare the test results equally between the various biometrics-based VPN systems.
    viii. Most importantly, as the CIO, document your test results from the initial testing of the biometrics-based VPN system and report any types of problems, malfunctions, and abnormalities to the vendors who have manufactured the requisite hardware and software so that the problem can be rectified and mitigated as quickly as possible.

Implementing a VPN

Now that we have looked at some of the major design implementations of a biometrics-based VPN system, the next step is its proper implementation at your place of business or organization.

Implementing your biometrics-based VPN system, you, as the CIO, can craft an easy four-step plan implementation plan:

1. Development of the VPN security plan
2. Preparation for the actual installation
3. Performance of the installation activities
4. The biometrics-based VPN system verification

In terms of the development of the security plan, you, as the CIO, need to carefully craft and design the policies, the procedures, and the rules that govern access to all of the network resources, which include the following:

1. The host computers and the servers
2. The computer workstations
3. The various network gateways, routers, firewalls, and network bridges
4. The terminal servers and the remote access servers
5. The network operating software
6. The operating system software of all of the servers
7. Any type of server application software that may be in use
8. All and any types of mission critical data found at your place of business or organization

However, keep in mind, as the CIO, building upon the security plan must include those end users who will be the most affected by the implementation of the biometrics-based VPN system. As the CIO, you need to carefully formulate the types of services that your end users will need from the system, especially getting permission rules for the user groups that are to be established.

Also, with regard to the VPN system rules for your end users, there must be extremely tight security, both physically (in terms of access control) and logically (granting whom has network access to the biometrics-based VPN system).

Equally important to the protection of the biometrics-based VPN system is that of the physical protection of both the public keys and the private keys. Also, as part of developing the security plan for the biometrics-based

VPN system, you need to collect information and data on the following types of hardware that exist in your IT infrastructure:

1. The external proxy servers
2. All of the internet routers
3. All types of firewalls at your place of business or organization
4. All of the virtual servers

In addition to the above, as the CIO, you will need to set up access control policies for the following items:

1. All of the VPN-based domain name service host names and the IP resolution-based tables
2. All of the e-mail network protocols used at your place of business or organization
3. The web servers in use (such as Internet Information Services for Microsoft environments and Apache for Linux environments)
4. Any Telnet and file transfer protocol networks and connections
5. Any other types of socket protocols and customized network protocols

Once all the above steps are accomplished, as the CIO, you then need to define the user groups that will exist at your place of business or organization. This step simply involves building and establishing end user tables as well as defining the access roles and rights of each group.

One key benefit of establishing these groups is that you can create a single entity that will fit all of the needs of each end user, rather than having to create rights and permissions for each individual employee at your place of business or organization.

The next phase is to properly define your biometrics-based VPN system tunnels, which include the following:

1. Remote user access to the biometrics-based VPN gateway
2. Gateway to gateway exchanges and connections
3. Gateway to remote partner exchanges and connections

Finally, you can get your biometrics-based VPN system set up for the final installation. This involves a full assessment of all of the following critical resources at your IT infrastructure:

1. Operating system readiness.
2. All dedicated hardware properly configured.

3. Physical access control checks to the VPN system are set in place and motion.
4. All of the network hardware and software are fully set to go.
5. All of the network routing tables to the biometrics-based VPN system are established and fully verified.
6. All of the host names at your place of business or organization match all of the domain name server entry names.
7. All of the TCP/IP addresses that relate to the biometrics-based VPN system are known, established, and operating properly and correctly.

Managing Public and Private Key Exchanges

One very important aspect to remember about installing a biometrics-based VPN system is that of key exchanges. Managing key exchanges can come in three different methodologies:

1. Manual Key Exchanges: With this system, a VPN administrator must configure each biometrics-based VPN system with its own set of public keys and private keys and also with the public keys and the private keys of the other biometrics-based VPN systems so that they can all communicate effectively and efficiently with each other. The prime disadvantage with this methodology is the lack of scalability.
2. The Simple Key Interchange Protocol (SKIP): This protocol is only available for highly commercialized applications and implementations of biometrics-based VPN systems.
3. The ISAKAMP/Oakley Protocol: ISAKAMP is the de facto standard for the IPv6 network VPN protocol and creates the framework for both the public key and the private key management. This type of VPN protocol does not actually the public keys and the private keys, rather it is used with the Oakley Network Protocol to produce and establish the public keys and the private keys for the biometrics-based VPN system.

As the CIO, a key principle to help fortify the security for the public key and private key management of your biometrics-based VPN system is to centrally administer the remote VPN Gateways and ensure that the interface to the biometrics-based VPN software is not available on any other network interface or device.

Digital certificates are a very important component of cryptography; moreover, they also play a very crucial role with the biometrics-based VPN system.

Since the biometrics-based VPN's primary function is to filter in and out the good and known data packets, these same types of digital certificates can be used to control access to all of the primary servers at your place of business or organization.

The Access Control List

This is done primarily via the ACLs. Essentially, this correctly identifies the list of approved end users along with their passwords at your place of business or organization.

Digital certificates can be used as a very strong proxy for each of the employee's username and password pairs. To make this happen, the digital certificates of each end user/employee associated with the ACL is mapped and can even determine and ascertain the access rights by reading each of the digital certificate's specific contents.

To realize the above process or your biometrics-based VPN system,

1. Have the ACL direct each new employee and end user to a brand new digital certificate for each separate site access.
2. The new employee/end user may have to select a new user ID/password combination.
3. The new employee/end user can then log back into the biometrics-based VPN system and the digital certificate is mapped to the new employee's/end user's user ID and password combination.

However, as the CIO, as the access rights of your new employees change, you surely would prefer to avoid the situation where new digital certificates have to be reissued each time the rights and privileges of your employees and end users change, per their job requirements and duties.

To do this, you, as the CIO, need to use an X.509 secret key to help secure your biometrics-based VPN system. To accomplish this task, follow this specific methodology:

1. Generate a random X.509 private key.
2. Because the digital certificate is already embedded into the biometrics-based VPN system, a challenge/response system is created.

3. Once the challenge/response has been successfully verified through the software at your IT infrastructure, it can then create its own challenge and automatically sign the end result, which will then use the private key associated with the X.509 digital certificate.
4. The signed result end user certificate is then sent back to the biometrics-based VPN server.
5. The verification of the end user's certificate is then performed.
6. If the digital certificate is proven to be valid, the end user's challenge is then checked, utilizing the public key in the employee's or end user's own digital certificate.
7. If the digital certificate is proven again to be valid, the resultant challenge is checked against the original challenge to ensure that they match.
8. If this particular challenge matches all of the previous steps just outlined above, the digital certificate number and information and data can then be extracted from the digital certificate number and information utilized as a lookup into the ACLs that contain those particular employee/end user access rights and privileges.

As one can see, as a result of the recent advent of the biometrics-based VPN systems, a number of key security drafts, over time, have been created around the security issues of VPNs, and these drafts are also known as the *Internet drafts*.

Internet Drafts

One such draft that relates to this chapter is known as the "Internet X.509 PKI." Specifically, this relates to the Elliptical Digital Certificate Algorithm (this is also known as the elliptical wave theory) in the PKI X.509 certificate schemes.

Here are the specific subjects, broadly speaking, this draft covers:

1. Certificate management messages over CMS
   This generally specifies the need for a certificate management protocol utilizing CMS in two different ways:
   a. The need for an interface that connects to the PKI products and services
   b. This outlines the needs for a certificate enrollment protocol using the Diffie-Hellman public keys.

2. The Internet X.509 Public Key Infrastructure Time Stamp Protocol
   Time stamps prove beyond a doubt that a transaction took place at a certain point in time, and thus it serves as a *trusted third party*, or TTP. This piece of the draft regulates how the format of the timestamp request should be sent to a time stamping authority and what type of particular response is valid.
3. The X.509 Public Key Infrastructure Data Certification Server Protocols
   This document describes very specifically the data certification services and the network protocols used when communicating with it:
   a. The validation of the digital signature
   b. Providing updated information with regard to the status of the public key certificates
4. The Internet X.509 Public Key Infrastructure PKIX Roadmap
   This document goes into the actual detail of the theory, and the implementation of the X.509 digital certificate-based PKI
5. The Internet X.509 Public Key Infrastructure-Qualified Certificates
   This document specifies the digital certificate profile for what is known as qualified digital certificates. The primary purpose of this document is to define the syntax of the language that shies away from all types of legal jargon.
6. The Diffie-Hellman Proof of Possession Mathematical Algorithms
   This specialized document details the particular of two separate methodologies to produce a digital signature from a Diffie-Hellman key pair. The mathematical algorithms are designed to create an audit trail rather than just for digital signing.
7. The X.509 Attribute Certificates
   This document specifies two very mathematical base profiles that are used to provide digital certificate authentication services
8. The Basic Representation Token V1
   This technical document helps to define data structures that have been established and distributed by a trusted provider
9. The Internet X.509 Public Key Infrastructure Extending Trust
   This technical document establishes the technical details that are needed to maintain a certain level of confidence by a trusted provider.
10. The Internet X.509 Public Key Infrastructure Operational Protocols, LDAP V.3
    This document highlights in details the LDAP V.3 features that are required to maintain a PKI-based upon the X.509 digital certificates.

11. The Simple Certificate Validation Protocols, also known as SCVP This specialized network protocol permits the client's computer to allow the primary server to handle the digital certificate processing. As a result, this type of server can then provide certain types of information and data about the details of the digital certificates, especially its validity or invalidity.

Four Vulnerabilities of a Biometrics VPN

With regard to the biometrics-based VPN, there are four critical areas where the data packets, which house the biometric templates, are at most risk to hacking and theft:

1. Just after template creation (this includes both the verification and the enrollment templates)
2. The biometric templates, which are housed in the database (the actual database depends upon the specific biometric technology being used)
3. In client–server network topology, the transmission of biometric templates from the biometric system to the central server (this is where the biometric database resides)
4. In a hosted environment, there the biometric template database resides with a third party

BIOCRYPTOGRAPHY

Biocryptography provides the means to further biometric templates at these critical junctures. Cryptography is the science of scrambling information and data that are in transit across a network medium and then descrambling it at the receiving end into a decipherable format.

That way, if the scrambled information and data were to be intercepted by a third party, there is not much that can be done unless they possess the keys for descrambling the information. These concepts of scrambling and descrambling can be very easily applied to biometrics. This is formally known as *biocryptography.*

In other words, the biometric templates are protected by scrambling and descrambling keys while they are stored in the database or in movement across a network. To further illustrate the exact nature of biometric templates, two types will be examined: fingerprint recognition and iris recognition templates.

To review, whenever we send a message to our intended recipient—whether by e-mail, instant message, or even just a text message on our smartphone, this message is often sent as a *plaintext* or *cleartext*.

This means that the actual message is being transmitted to the intended recipient in the way it was originally constructed by the originator of the message. Thus, the only true way to protect the information being sent is to scramble it, in other words, *encrypt the message*. This encrypted (or now scrambled) message is now known as the *ciphertext*.

The reverse of this process is known as *decryption*, with the end result being a readable message to the intended recipient. As all of this relates to biometrics, the data packet that houses the biometric template (such as the fingerprint or iris recognition template) can be viewed as the plaintext or as the *plaintext biometric template*.

The Cipher Biometric Template

When the fingerprint or iris template is encrypted, it can be viewed as the *cipher biometric template*, and when it is decrypted, it can be viewed again as the decrypted *plaintext biometric template*. However, other than just doing the above, biocryptography also has to provide the following three functions for it to be truly effective:

1. *Authentication*: The receiver of the message (or the plaintext biometric template) should be able to 100% verify its origin.
2. *Integrity*: The message in transit (or the plaintext biometric template) should not be modified in any way or format while it is in transit (or in other words, replacing a fingerprint biometric template with an iris biometric template to spoof the biometric system).
3. *Nonrepudiation*: The sender of the plaintext biometric template should not falsely deny that they have not sent that particular template originally.

Biocryptography Keys

A component that is central to biocryptography is what is known as *keys*. It is the key itself that is used to lock up the plaintext biometric template (or encrypt it) at the point of origination, and it is also used to unlock that same template at the receiving end.

The key itself is a series of mathematical values—as the value becomes larger, it becomes equally hard (if not harder) to break while in transit. The range of possible mathematical values is referred to as the *keyspace*.

There are many types of such keys used in biocryptography, such as signing keys, authentication keys, data encryption keys, session keys, etc. The number of keys that are generated depends primarily upon the mathematical algorithms that are used, which are primarily symmetric and asymmetric mathematical algorithms.

To further fortify the strengths of a biocryptography-based PKI, mathematical hashing functions are also used to protect the integrity of the plaintext biometric template. For example, when the destination party receives the plaintext biometric template, the hashing function is included with it.

If the values within the hashing function have not changed after it has been computed by the receiving end, then one can be assured that the plaintext biometric template has not been changed or altered in any way.

To prove the validity of the hashing functions, it should be noted that they can be calculated only in one direction (e.g., going from the sending point to the receiving point, where it is computed) but not in the other direction (e.g., going from the destination point to the origination point).

A Review of How Biocryptography Can Be Used to Further Protect Fingerprint and Iris Templates

As discussed in Chapter 3, biometric systems can take quite an array of system designs, deployments, and network architectures. To illustrate that the concepts of cryptography can be used with biometrics, three primary examples will be reviewed:

1. From the standpoint of a single biometric system
2. A client–server setting, where the biometric devices are connected to a central server
3. A hosted environment, where the biometric templates and the processing functions are placed in the hands of a third party

Biocryptography in a Single Biometric System

With the biometric technology that is available today, most fingerprint recognition and iris recognition scanners consist of the database and the processing functions in just one unit, i.e., enrollment and verification occur at a single point.

Obviously, there are many advantages to having this type of *stand-alone* system, with the two biggest ones being low overhead in terms of costs and very quick processing times of the enrollment and verification templates.

This is how biocryptography would be used to protect the iris and fingerprint recognition templates in this type of environment:

1. Assuming that the end user wishes to gain physical access or logical access with either their fingerprint or their iris scan, a verification template must be first created. This template would either be a binary mathematical file (for a fingerprint template) or an IrisCode (for an iris template).
2. Using the principles of symmetric cryptography, the verification templates would then be encrypted with the key that is generated by the system. This would occur right after the unique features from the fingerprint or the iris are extracted, and the template is created.
3. Once the verification template reaches the level of the database, it would then be decrypted by the same key, and the statistical correlations would then be computed between the verification and the enrollment templates. If this correlation is within the bounds of the security threshold established, then either physical access or logical access would be granted by the biometric system.

In this illustration of the stand-alone iris or fingerprint recognition system, a number of key assumptions are made:

1. Only verification or a 1:1 match is being used. Since it is being done at the local level, the configuration needs are low, thus symmetric algorithms are the key choice of cryptography to be used. As a result, only one key is generated.
2. Only the verification templates receive the added protection from encryption. This is because the fingerprint and iris recognition templates are created only once and are later discarded from the respective biometric system. In fact, in any biometric system, no matter what the magnitude of the application is, verification templates are used only once, and no more.
3. The enrollment templates in the fingerprint and iris biometric system receive no added protection from encryption. There is no

doubt that this is an inherent security risk, but one has to keep in mind also that the database is being stored in the biometric device and not located in multiple places, where the need for protection is much greater.

4. In a biometric verification application at the local level, the processing power required is much less than when compared with the client–server or even the hosted approach. Thus, this further supports the need for only the use of symmetric algorithms.

Biocryptography in a Client–Server Biometric System

In a client–server network topology in the traditional sense, there are a series of computers connected to a central server via a network medium (for example, it could be a hard-wired network or even wireless).

Within this central server resides all of the resources and applications the end user needs access to. Not only are databases often stored here, which contain all types of data, but this is also the point where database querying and processing takes place.

This type of infrastructure can be very small (also known as a local area network, or LAN), or it can be very large, covering great distances and international boundaries (this is known as a wide area network, or WAN).

This type of setup can also be extrapolated to biometrics. Multiple biometric devices can be networked to a central server, in a very similar fashion as described above. However, the key difference in a biometrics client–server system is that the primary application is for verification and identification, template processing/querying, and nothing more.

Currently, biometric client–server applications exist in business. Typically, it is the medium- to larger-sized businesses that have this type of configuration because many more resources are required, which can be a much greater expense compared with the single biometric system discussed previously.

In this type of configuration, the most commonly used biometric devices are that of hand geometry scanners, fingerprint scanners, iris scanners, and even facial recognition scanners. Different biometric devices can be used (such as fingerprint scanners being used in conjunction with iris scanners) or the same biometric devices (for example, all fingerprint scanners).

In the end, it really does not matter what hardware is used because they are all accessing the same resource—namely, the central server. However, with all of this comes an even greater risk—specifically the security threats that are posed to the system.

Again, biocryptography can play a huge part to ensure the protection of the biometric templates. For this configuration, asymmetric cryptography will be used. Meaning, a public key will be generated as well as a private key.

This is how biocryptography will be used in this type of environment:

1. The end user either has his or her finger or iris scanned, and the usual verification templates are created by each biometric system (assuming that both iris and fingerprint scanners are being used simultaneously).
2. These iris and fingerprint verification templates will then be protected by a key, specifically, the public key.
3. After the above two processes have been completed, the newly encrypted verification templates will then make their way across the appropriate network media, and finally make their way to the central server.
4. It is here where the biometrics database is stored, which contains the relevant iris and fingerprint enrollment templates, and it is here where the private keys are stored and where these templates are encrypted.
5. Once the public and private keys have been decrypted, the appropriate statistical measures will be applied to determine the closeness or the dissimilarity between the verification and the enrollment templates.
6. Based upon the results, the end user will be granted or denied access to whatever application they are trying to gain access to (either physical access entry or logical access).

An important point needs to be made about this biometrics client–server system: As is the case in the single biometric system, the verification templates, after they have been compared and evaluated, will be discarded along with their public keys. The enrollment templates will have to be decrypted as well, so that the comparison between the verification and the enrollment templates can be made. However, because the enrollment templates are stored from outside the actual biometric device, they will have to be re-encrypted again to ensure maximum security while they are stored in the database at the central server. Thus, this type of configuration will need far more network resources and much more processing power. Also, there is no doubt that extra overhead and quite possibly further verification times will increase by a few more seconds (under normal conditions, this happens in less than 1 second).

Biocryptography in a Hosted Biometrics Environment

The world of IT is now moving very quickly toward a new type of application—which is known as *cloud computing* or *software as a service* (SaaS). This means that a businesses' entire IT infrastructure can be managed and outsourced to an independent, third party, also known as a *hosting provider.*

At this venue, all of the IT infrastructure hardware and software is set up and managed, maintained, and upgraded by them. All a business owner has to do is open an account with this hosting provider, and with a few clicks of the mouse, set up the IT services they need or desire to have.

The primary advantages of this are (1) total elimination of IT administrative problems because the hosting party is held entirely responsible for everything and (2) a business only pays for the software/hardware services they have subscribed to at a fixed monthly cost.

This model of cloud computing can now even be expanded to the world of biometrics, which can be termed as *biometrics as a service* (BaaS). This is also known as a *hosted biometrics environment*. It would be established as follows:

1. All that a business would have to do is purchase and acquire the requisite biometrics hardware, in this scenario, it would be the fingerprint scanners and iris scanners.
2. All of the servers, the databases that house the iris and fingerprint enrollment templates, the processing of the verification templates between the enrollment templates, and the formulation of the match/nonmatch result rest entirely with the hosting provider, i.e., after the biometrics hardware is installed, all the business has to do is simply set up the services needed with their account, and all is set to go.

However, because the iris and fingerprint templates are placed in the hands of the hosting provider, security of these templates obviously becomes the prime concern.

Biocryptography and VPNs

There is yet another tool of cryptography that would be perfect here—VPN. This is how it works with a BaaS type of application:

1. The end user in the previous examples of this article has his or her fingerprint or iris scanned to create the respective verification template. This template then gets broken down into a separate data

packet. This data packet (which contains the iris or fingerprint verification template) is then further encapsulated (or encrypted) into another data packet, so it eventually becomes invisible as it traverses across the various network media, as it makes its way to the servers of the hosting provider.
2. To ensure the integrity of this double-layered data packet, it would also consist of headers that contain information about the size and type of the verification template. This would be a confirmation to the hosting provider, as they receive this data packet, that none of the iris or fingerprint template data have been altered or changed en route.
3. To create another layer of protection, a dedicated VPN channel can be created. This would be a direct line from the point of origin of the fingerprint or iris scanner all the way to the servers of the hosting party. This is known specifically as *IP tunneling*, and as a result, this channel cannot be seen by other people accessing the Internet across the same network media the data packets (which contains the verification templates) are also traveling across.
4. Once this data packet arrives at the servers of the hosting provider, it is then decrypted, and the same processes of verification/enrollment template comparing, as detailed in the biometric client–server network system, will be followed.
5. After a match or nonmatch has been determined, the result is then sent from the hosting provider back to the place of origin of the iris or fingerprint scanner, and the end user is then allowed or not allowed access to the resources or applications they have requested.

IPSec

To further fortify the VPN between the place of business and the hosting provider, a protocol known as *IPSec* can be used. The IPSec protocol is a major security enhancement to the TCP/IP protocols currently used to gain access to the Internet.

IPSec also uses digital certificates (which are also the public and the private keys). It should be noted that there is one specific type of IPSec mode that helps provide the maximum security possible to the data packet housing the verification template.

It is known as *IPSec tunneling*, in which the header and the payload it is carrying (which is again, the iris and the fingerprint verification template) are further encrypted at a much deeper level.

Some special points need to be made about BaaS:

1. Although this type of application is still very new (only voice biometrics thus far has been used as a BaaS), it could very well be the wave of the future. This is so because biometrics as a security technology is still perceived to be very expensive. However, with a hosted approach, these high costs will be greatly eliminated because all that is required of the business owner to purchase is the required biometric hardware, and also, like SaaS, BaaS would be a fixed monthly cost.
2. With BaaS, the verification times and presenting the match/nonmatch result could take time because all of the template processing and matching will take place at the hosting provider. The speed will be a direct function of the hardware and the software that is used as well as the amount of network bandwidth that is consumed.
3. Although BaaS holds great promise, one of the biggest obstacles that it will face is one that has constantly plagued the entire biometric industry, which is the issue of privacy rights. This issue will only be proliferated more as the biometric templates are in the hands of an outside third party, which is the hosting provider.

In summary, this chapter has examined how the principles of cryptography and biometrics can be used together to provide maximum security for the most important component, namely the biometric templates that are created and stored. As of this writing, biocryptography is still very much an emerging field, and some observations have to be noted:

1. The three examples described and discussed in this article are theoretical in nature—they have not yet been proven in the real world, and it is assumed that much research will have to be conducted before these applications are put to the test on a commercial basis.
2. These examples, if proven to be viable in the real world, could become very complex, very quickly. This is so because there are two types of security technology becoming fused as one, and from that, many variations of all kinds and types could be

created. Thus, great importance has to be paid to the actual biometric systems analysis and design right from the outset to help ensure a smooth streamlined process, from the standpoint of troubleshooting and support.

3. Another area that has haunted the biometrics industry is that of a lack or sheer absence of standards and best practices for the technology both in its current state and as it is being developed. As biocryptography emerges into the forefront of security, this is an absolute must, in order to avoid duplication of efforts, resulting in unneeded and bloated overhead. Also, as businesses and entities start to adopt biocryptography, a standards and best practices list will help to provide the groundwork that is needed to create security applications, rather than having to reinvent the wheel every time.
4. Finally, it is the author's point of view that biocryptography be developed in an open model type of forum, where all parties involved, ranging from the private sector, to academia, to the level of the government, can collaborate and discuss new ideas with the goal of open communications and open dialogue. An open model would help to minimize any potential security threats and risks to biocryptography because answers and solutions can be very quickly created. This is best exemplified by the use of the open-source model for software development versus the closed source model.

REVIEW OF CHAPTER 4

The previous chapter looked at and reviewed the emerging field of biocryptography in extensive detail. An extensive overview of cryptography was provided as well as how biometrics could very easily fit into this scheme to help further fortify and strengthen the biometric templates that are created. In summary, some of the major components reviewed were symmetric and asymmetric cryptography, the components of PKI (also known as PKI), as well as VPNs.

As was demonstrated in the last chapter, it appears that VPNs will be the way to go in the world of biometric technology. Essentially, this involves encapsulating the data packet where the biometric template resides in with another data packet. This particular data packet then traverses across a network connection that is virtually invisible to the outside world.

As it was also reviewed, all of the mathematical algorithms involved in biocryptography were examined, and it appears that the RSA algorithm will provide the most robustness for biocryptography, in a virtual private setting. This is so because the RSA algorithm relies upon the power of very large prime numbers to *crack its code.*

5

An Introduction to Biometrics in the Cloud

INTRODUCTION TO CLOUD COMPUTING

In this chapter, we cover a very important area where biometrics will play a role again—cloud computing. Cloud computing has become a very popular phrase in the twenty-first century, especially to that of corporate America. To many people, the only thing that is truly understood about it is that it brings much lower, fixed costs to a business owner at a predictable price. Also, it is popularly understood that an entire IT infrastructure can be placed at the hands of a third party, known as the Internet service provider (ISP).

However, truth be told, there is far more to cloud computing than just the above. In fact, cloud computing is still a very nebulous concept. Therefore, the goal of this last chapter is to dispel myths surrounding cloud computing and provide the CIO with a strong background into it, and above all, how biometrics can fit into and play a role within the cloud-computing infrastructure. Therefore, this last chapter will cover the following topics:

1. Basic concepts and terminology surrounding cloud computing
2. Benefits of cloud computing
3. Challenges and risks of cloud computing
4. Functions and characteristics of cloud computing
5. Cloud-computing delivery models
6. Cloud-computing deployment models

7. Security threats posed to cloud computing
8. Important mechanisms of cloud computing
9. Cloud computing cost metrics and service quality mechanisms
10. An introduction to biometrics in the cloud
11. A detailed example of how biometrics in the cloud will look like
12. The advantages and the disadvantages of biometrics in the cloud

THE BASIC CONCEPTS AND TERMINOLOGY SURROUNDING CLOUD COMPUTING

When one thinks of cloud computing, especially the business owner, people get very excited. Why is this so? As alluded to in the introduction of this chapter, it is the buzzword that is floating around today and will be for quite some time to come. Many business owners feel that if they adopt a cloud-computing infrastructure, they will be ahead of their competitors. Although this may be true to a certain extent, the chances are that the competition has already beaten them to the punch.

This is because, in theory at least, cloud computing can be configured and started with just a few clicks of the mouse and at the fraction of the cost of owning an entire IT infrastructure. Before we go any further into the components of cloud computing, what exactly is cloud computing?

According to the National Institutes of Standards and Technology (NIS), cloud computing can be defined as "a model for enabling ubiquitous, convenient, hands on, on demand network access to a shared pool of configurable computing resources (e.g., networks, servers, storage, applications, and services) that can be rapidly provisioned and released with minimal management effort or service provider interaction. This cloud model is composed of five essential characteristics, three service models, and four deployment models" (*Cloud Computing: Concepts, Technology and Architecture*, Erl, T., 2013, Arcitura Education, p. 28).

This no doubt seems like a long and complicated definition; we now provide a much more distilled definition of cloud computing: "cloud computing is a specialized form of distributed computing that introduces utilization models for remotely provisioning scalable and measured resources" (*Cloud Computing: Concepts, Technology and Architecture*, Erl, T., 2013, Arcitura Education, p. 28) (an example of cloud computing can be seen in the following figure).

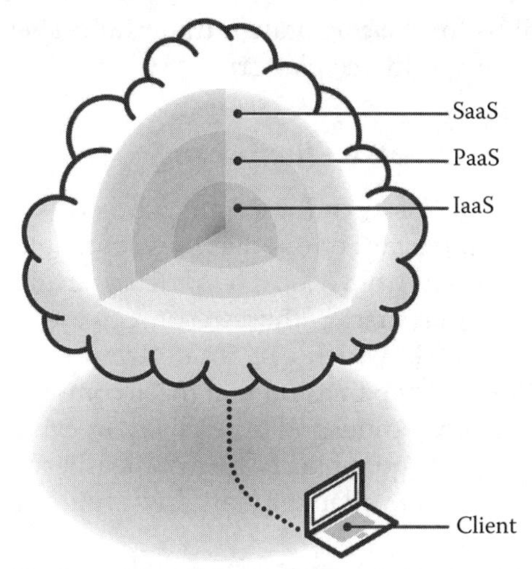

Now, as we take both of these definitions, we find that cloud computing consists of a number of key components, which include the following:

1. The cloud
2. The IT resource
3. On-premise
4. Scalability

THE CLOUD

As we progress through these major components, the first one we look at is the cloud itself. This component can be specifically referred to as a "distinct IT environment that is designed for the purpose of remotely provisioning scalable and measured IT resources" (*Cloud Computing: Concepts, Technology and Architecture*, Erl, T., 2013, Arcitura Education, p. 28). As we explore this component, the cloud is like a nebula in which various types of computing resources can be accessed.

These resources include the entire gamut of computing applications, which ranges from the database to the individual software packages that service an entire small business, to accessing the file transfer protocol server, to accessing your e-mail server, to even having the capability to develop and create complex software applications, to accessing content management systems for the corporate intranet, and to even creating and launching an entire e-commerce platform.

Two Distinctions

However, at this point, it is very important that two key distinctions be made. The first very important one is that these computing resources just described are not accessed from a local hard drive on an employee's workstation or laptop. For that matter, these resources are not even accessed from a local area network, which can reside from inside a corporation. Rather, it is important to keep in mind that these computing resources can be accessed from literally a thousand miles away, or even across the globe. The remote access is made possible because of the Internet. With any standard web browser that is available today, any type of cloud computing resource can be accessed.

The second very important distinction to be made is that all of these computing resources reside within an independent third party, which is privately owned, such as that of ISP.

The IT Resource

The second major component we now examine is that of the typical IT resource, which can be found from within a cloud-computing infrastructure. An IT resource, or even an IT asset, can be defined "as a physical or virtual IT-related artifact that can be either software-based, such as a virtual server or a custom software program, or hardware-based such as a physical server or a network device" (*Cloud Computing: Concepts, Technology and Architecture*, Erl, T., 2013, Arcitura Education, p. 34).

Based upon this definition, an IT resource/IT asset is a tangible item that can be placed into the cloud, even some of the software applications that were just previously described. However, in the world of cloud computing, the IT resource/IT asset is typically a server, whether it is a standalone physical server or just a virtual server. A virtual server is also a physical server, but is partitioned so that each of the resources it hosts looks like it resides in a separate server to the end user or small-business owner.

Thus, as you can see, the most important IT resource, based upon the definition, is that of the server. After all, the software applications just previously illustrated need a place to reside in, or in cloud computing terminology, need a place to be *hosted* at. However, a key distinction must be made here. Although the virtual server can be placed into the cloud—it also represents all of the resources that are available to the end user.

A key feature is that cloud computing relies heavily upon shared resources to bring the low and predictable monthly costs to the end user and the small-business owner.

On Premise

The third major aspect to be examined in the cloud-computing infrastructure is that of the *on-premise* component. Specifically, on-premise can be defined as "being on the premises of a controlled IT environment which is not cloud-based" (*Cloud Computing: Concepts, Technology and Architecture,* Erl, T., 2013, Arcitura Education, p. 36). A key part of this definition is that of the *controlled IT environment.* This environment is actually that of the ISP, or for that matter, any other private entity that has the flexibility to provide at least one or more IT assets and resources.

Scalability

An important detail that needs to be made at this point is that the cloud-based resources are not just stored onto one specific computer or even just one particular server; rather all of those IT assets and resources are shared across hundreds, or even thousands, of servers at ISPs all over the world.

As we are now starting to see emergence, the cloud consists of a controlled environment that is that of the ISP. The ISP consists of the IT resources and the IT assets that reside on a virtual server, from which these particular resources and assets can be accessed by the end user or the small-business owner.

This is just one example of a general cloud-computing infrastructure, and there is not just one of them. There are hundreds and thousands of them worldwide to bring to the end user and the small-business owner the shared computing resources they need and require. As will emerge throughout the rest of this chapter, the cloud possesses many strong benefits, which are available to the end user and small-business owner at an extremely affordable price, compared with the other IT assets and resources that are available in the traditional models.

This will be especially true when biometrics becomes more available via the cloud infrastructure. On the spectrum of security technologies that are available today, biometrics technology tends to be more on the expensive side in terms of the customer with regard to procurement and deployment. However, with the cloud resources that are available today and well into the future, an enterprise grade-level biometrics technology infrastructure will soon be able to be procured and deployed by even the smallest of businesses and organizations, which at the current time can only be afforded by the large corporations.

However, apart from costs, the cloud infrastructure possesses other key benefits such as

1. IT asset scaling
2. Greatly reduced investments and proportional costs
3. Scalability
4. Increased IT availability

Asset Scaling

In terms of the first major benefit, IT asset scaling merely refers to the fact that an IT asset or even IT resource can handle increased or decreased end user demand usage. There are two types of scaling available:

1. Horizontal scaling
2. Vertical scaling

With horizontal scaling, this is the allocation or the provisioning of valuable IT assets and resources that are of the exact same nature and type. The allocation of these particular IT assets and resources is known as *scaling out*, or the releasing, and the giving away of IT resources to newer end users and newer small-business owners is known as *scaling in*.

For example, let us illustrate this with a well-known application from Microsoft, known as the SQL Server. This database software is widely available through most ISPs (also known as *cloud providers* in some circles). The allocation of the SQL Server to the existing end users of this particular application is known as scaling out. However, the releasing of the SQL Server application to the new end users or the new small-business owners when they first sign up for cloud-based services is a perfect example of scaling in. This type of scenario is very typical among the ISPs who provide cloud-based IT services.

With regard to vertical scaling, the valuable IT assets and resources are merely replaced with a lower or higher version of the very exact same type of IT asset or IT resource. For instance, replacing an IT resource or an IT asset with a higher version is known as *scaling up*, and the opposite, which is the replacement of an IT asset or IT resource of a lower version is known as *scaling down*.

Let us illustrate again with the same SQL Server example, but this time, with a known version of SQL Server 2012. The offering to the end user of a lower version of SQL Server (such as SQL Server 2000) is known as scaling down, and the offering to an end user of a later version of SQL Server 2012 is known as scaling up. This type of scaling is less commonly offered by the ISPs as opposed to the horizontal scaling.

This is so because vertical scaling requires more down time. Vertical scaling typically only happens in the cloud environment when a later version of an IT asset or IT resource becomes available and makes the present IT asset or IT asset totally obsolete and outdated. With respect to the second major benefit of cloud computing, which is that of reduced investments and proportional costs, it is quite true that many of the ISPs who offer cloud computing services to their end users can have a revenue model based upon the bulk acquirement and acquisition of the IT assets and resources, and in turn, offer these services to their end user customer base at very cheap price point packages.

This is what directly allows the owners of small- to medium-sized businesses to purchase a Fortune 100 enterprise grade-level IT infrastructure for virtually pennies on the dollar. This is obviously of great appeal to small-business owners, as they can now replace capital operational costs and expenditures (which also means the total investment and ownership in an outright, entire IT infrastructure) with a proportional cost expenditure (which means the rental in a cloud-based IT infrastructure).

Because of the benefits provided by the proportional cost expenditures, this permits the small- to medium-sized enterprises to reinvest the saved money into other much needed IT assets and resources to support mission critical business functions.

However, keep in mind that ISPs are also businesses who have revenue generating models, and to afford such mass acquisition of IT assets and resources, they need to find inexpensive alternatives also. Thus, the ISP locates both types in data centers where the two most important cost variables of network bandwidth and real estate are the cheapest possible.

Proportional Costs

In this second category of a major benefit, there are a number of sub-benefits of cloud computing that are also available to end users and small-business owners:

1. The access to on-demand IT assets and resources are a pay-as-you-go basis, and the ability to reallocate these resources to other end users who are willing to pay for them when they are no longer needed by existing end users and small-business owners.
2. Having the feeling that the end user and the small-business owner have unlimited use to IT assets and resources.
3. The ability to quickly add or subtract/delete any type of kind of IT resource/IT asset at the most granular level (such as modifying an end user's mailbox through the e-mail interface).
4. With cloud computing resources, the IT assets and the IT resources are not locked into any physical location so the IT assets and the IT resources can be moved around as needed.

However, despite these benefits, it can be quite a complex financial process for a Fortune 100 company to decide if they wish to have their own on-premises IT infrastructure versus provisioning cloud-based resources, as opposed to the small- to medium-sized business, where the decision is so obvious and clear cut.

Scalability

With regard to the third benefit, that of scalability, this simply refers to the ability of an IT resource or an IT asset to dynamically meet the needs of the end user or small-business owner. In most types of scenarios, this dynamic nature of the IT asset or IT resource can take quite a bit of time to adjust accordingly.

However, when it comes to the cloud-computing infrastructure, these particular IT assets and resources have to adjust to the need of the end user or the small-business owner in just a matter of seconds or less once their cloud-computing infrastructure account has been configured and the desired IT assets and resources have been selected and paid for.

This scalability feature of the cloud and its dynamic nature to meet the needs of the end user and small-business owner are among the biggest benefits of the cloud-computing infrastructure and among its strongest selling points to the small- to medium-sized business. Imagine a small- to medium-sized business that possesses an in-house IT infrastructure. The nature and the scope of the business has grown in size, as well as its usefulness and scope. Consequently, the cost of upgrading the entire IT infrastructure or even just various components of it can be very cost prohibitive for that particular small- to medium-sized business.

However, if they had adapted their IT assets and resources through a cloud-based infrastructure, they could have brought in the new services needed in just a matter of seconds, and of course, at just a fraction of the cost. Also, it is not just a matter of meeting the specific needs of the end user of the small-business owner; the IT resources and the IT assets that are founded and based in the cloud-computing infrastructure can also dynamically meet the needs of the processing fluctuations (of the IT assets and the IT resources) as required.

For example, if an end user requires more disk space, he or she can merely increase it in just 2 s with a few clicks of the mouse. However, with a traditional in-house IT infrastructure, the entire, physical hard drive would have to be replaced, thus causing the end user an extra expense. This dynamic scalability of the cloud-computing infrastructure is a direct function of a concept introduced before, known as proportional costs.

The end user or the small-business owner pays for those IT assets or IT resources they specifically use, no more and no less. As a result, the cloud-computing infrastructure has adapted the term *as a service*.

Availability and Reliability

One of the last major benefits of the cloud to be reviewed is that of increased availability and reliability. Translated to the language of the small-business owner, this simply means that all of the IT assets and resources within a specific cloud-computing infrastructure is *always going to be on*. One may ask at this point how an ISP can provide this level of availability to end users and small-business owners? Think about the actual infrastructure that the ISP resides in.

It is a full-blown data center, with dozens of redundant servers, power lines, and even redundant power supply backups, to provide a continual supply of power to the data center, so that its IT resource and IT assets will always be available and online to end users and small-business owners 24 × 7 × 365. However, despite all of these redundancies and the best efforts that are afforded by the data center, downtimes do occasionally happen.

In these particular cases, remedies have to be provided to the end user and the small-business owner. Another beauty of the cloud-computing infrastructure is that even as a particular cloud segment grows and expands, the addition of newer IT assets and resources and the deletion of older IT assets and resources, interruption of the cloud-based services to the end user or small-business owner hardly ever happens. It is as if *business as usual* is happening.

SLA Agreements

All of these uptimes that are guaranteed by the ISP come to the level of the written contract known as the *service level agreement*, or SLA. In this contract, the ISP specifies to what levels of uptime it can guarantee to the end users and small-business owners. Very often, this uptime is defined as the *five 9's*, which means that 99.999% of uptime will be made available to the end user or the small-business owner, and when this level is not reached, the ISP has to provide remediation to its customer base.

Therefore, it is very important to the customer, whether it is an end user or a small-business owner, to read this contract carefully before signing it.

The Challenges and Risks of Cloud Computing

Along with these perceived benefits of the cloud, unfortunately, come its disadvantages as well. Remember, the cloud is still a relatively new concept, and the boundary lines it possesses are still not clear. For example, a small-business owner's cloud-computing infrastructure could actually be shared with another cloud, depending upon the provisioning of the various IT assets and resources.

Naturally, there is also the trust factor when placing your entire IT infrastructure into the hands of the ISP, which in this case, could be halfway around the world, to people you have never met before, let alone even heard of. To make matters worse, you are even trusting your credit card information with this third party, whom again you do not even know. You are even sharing and depending your businesses' confidential and proprietary data and storing them into the cloud, with other cloud resources.

In this regard, privacy may not be assured. In addition, there is always the threat of break-ins by hackers who can steal all of your critical data. However, despite all of this, the large-scale thefts of confidential and proprietary information and data are rare, but it is very important that you, as the end user or the small-business owner, understand these leading, inherent risks, which can be classified as follows:

1. Increased security vulnerabilities
2. Reduced operational governance control
3. Limited portability between the cloud processes
4. Multiregional compliance and legal issues

Security Risk and Challenge

In terms of the first security risk and challenge, which is about specific security vulnerabilities, the moving of private and confidential business data ultimately becomes the responsibility of the ISP with whom the end user or the small business has entrusted the data. True, the contracts and the SLA agreements have been signed, but it still takes quite a bit of faith for the small-business owner to transfer all of his or her corporate information and data to a yet untrusted third party.

This is probably one of the biggest fears that the ISP has to allay. Given the wide expanse of the cloud-computing infrastructure, access to IT assets and resources cannot be given directly to the end user or the small-business owner. Rather, these specific IT assets and resources have to be accessed remotely (such as via file transfer protocol, or FTP), which can pose even greater threats and risks, especially to a hacker who is *listening* in on the other end.

This means that all of the IT assets and resources that reside from within the cloud-computing infrastructure have to be reached externally via various access points. The bottom line is that there is no 100% fool-proof way to secure those particular access points to the cloud-computing infrastructure.

Another major security threat is that of access to confidential consumer data that may be stored in a cloud-computing infrastructure, for whatever reason (for example, a retail business owner may be using their cloud infrastructure to back up their customer information and data). Data security and information safety and reliability of this information rely solely upon the security controls and policies implemented by the ISP.

Finally, another grave security threat or risk posed to the cloud-computing infrastructure is the intersection of cloud boundaries and the malicious cloud consumers it can create. For example, with shared IT resources and shared IT assets, there are much greater dangers for the theft and damage of mission critical business data stored in the cloud.

The bottom line is that it can be very difficult for an ISP to provide strong security mechanisms when cloud boundaries intersect one another. This cloud-computing infrastructure threat is also known as *overlapping trust boundaries.*

Reduced Operational Governance

With respect to the second major security threat category, which is that of reduced operational governance, control simply refers to the fact that end users and small-business owners who use the cloud finally have some control over the IT assets and resources that they own or rent. For example, imagine an end user at his or her place of business or organization. He or she once had to abide by strict IT security policies. However, now, this end user has newfound freedoms to fully control the IT assets his or her business owns or rents in the cloud-computing infrastructure.

This feeling of unfounded power that besets the end user can translate into real risks for the ISP into how it should control its cloud-computing infrastructures. Now, most end users may get a short-term rush on this new control they now possess, but these are the end users whose rush can extend into a hacker's mindset and try to gain access to other end user's clouds, and thus to hijack their information and data.

This puts the ISP at graver risk, as well as the lines of communication that exist between the cloud end users and their own cloud-computing infrastructures. Two distinct consequences can arise from this: (1) the guarantees as established by the SLA can become null and void, thus causing the ISP significant financial loss and risk; and (2) bandwidth constraints can occur when the cloud-computing infrastructures are located at much further geographic distances from within the platform of the ISP.

One of the best ways, and probably the only way, to avoid these kinds of risks and threats is for the ISP to keep a close and vigilant eye on all of the IT assets and resources that are being used by all of the cloud users. Any suspicious activity or tampering by would-be hackers in the cloud-computing infrastructure must be dealt with promptly and deftly by the ISP.

Limited Portability

The third major security risk with a cloud-computing infrastructure is that of limited portability, which is available among all of the ISPs. Essentially, what this means is that with all of the ISPs around, there is no common set of standards or best practices overall.

Meaning, each ISP can carry its own set of particular technologies at various version levels. Although there is nothing inherently wrong with this approach, it can cause major security concerns for the end user and the small-business owner, who is wishing to switch to a different ISP. For example, if an end user wishes to transmit and switch over their customer database from one platform to another cloud computing platform at the new ISP, there is no guarantee that this confidential and proprietary information and data will transfer over to the new cloud computing platform cleanly and easily. In fact, some of these data could even be hijacked by a third party, such as a hacker.

A lot of this reason has to do with the fact that many cloud computing architectures and platforms are, to a certain degree, proprietary from one ISP to another. Let us look back at our previous example.

Suppose that the small-business owner has a customer database that is custom made and is dependent upon other various technology platforms. The new ISP may flaunt serious claims that this customer database will transfer smoothly over to their own proprietary cloud architecture. However, the truth of the matter is, what if this customer database does not transfer cleanly from the existing cloud-computing infrastructure to the new cloud-computing infrastructure?

Although the small-business owner could be blamed for not doing their part of due diligence, i.e., to analyze the technological compatibilities for this particular database, the real accountability will rest with the new ISP. What lacks between these ISPs is a set of best practices and standards that allow for the ease of transfer of information and data for the end user from one cloud-computing infrastructure to another ISP.

However, at least for now, the ISPs are starting to realize this big gap and, in lieu of it, are now offering cloud computing technologies, which are standard among the industry. Having this level of standardization is especially important for the biometric templates, especially when they are stored in a cloud-computing environment.

With regard to the biometric templates, any loss of any type of magnitude from the transfer of one cloud-computing infrastructure to another cloud-computing infrastructure will have far-reaching legal ramifications in both the court of law and the judicial processes.

Compliance and Legal Issues

The fourth major security vulnerability posed to a cloud-computing infrastructure is that of the multiregional compliance and legal issues that can arise. What does this exactly mean? The cloud-computing infrastructure, or any matter related to it, is located in different geographic regions around the world.

For example, although a small business or organization might be based in the United States, the owner of that particular business or organization can purchase a cloud-computing infrastructure from halfway around the world, such as Germany or Russia. One of the advantages is the very cheap price that can be offered to the small-business owner. However, to the small-business owner, it makes no difference where the actual, physical servers are located from within the cloud-computing infrastructure.

Although most of the time this is not the case, the information and data housed in a cloud-computing infrastructure in a different country could very well be subject to the laws of the country where the infrastructure is. To this effect, these foreign governments could very well improve their own data privacy and storage policies.

Another strong security concern is that of the accessibility and disclosure of the information and data that reside in the cloud-computing infrastructure. For example, if a small-business owner in Europe owns a cloud-computing infrastructure in the United States, their information and data could be prone to inspection by the U.S. government based upon the guises of the U.S. Patriot Act. This is one of the weakest areas of putting a biometric system into a cloud-computing infrastructure. This is because the storage of all of the biometric templates on a cloud-computing infrastructure is a very sensitive topic by itself, and if this cloud-based biometric database does come under heavy scrutiny by a government agency, cries of violations of civil liberties and privacy rights will abound, thus defeating the purpose of a biometrics cloud-based infrastructure.

However, this also brings up the issue of whether a biometrics cloud-based infrastructure should be stored in a public cloud versus a private cloud.

The Functions and Characteristics of Cloud Computing

Thus far, we have reviewed in this chapter the definition of cloud computing, its major components, as well as its benefits and the major risks in terms of security that are posed to a cloud-computing infrastructure. Now, in this part of the chapter, we turn our attention over to the characteristics that separate the cloud-computing infrastructures from the actual IT infrastructures that exist today.

As we have alluded to throughout this chapter, the cloud, to many people, is a very nebulous concept. Nebulous in this case means very murky, with no clear boundaries or definitions. True, in many aspects, this is what the cloud-computing infrastructure is about. It comes from out of somewhere, in some part of the United States, or for that matter, in any part of the world, even in the most remote region imaginable.

All one has to have is a computer, an Internet connection, and literally from thousands and thousands of miles away, the access to an ISP. With these, all the end user or the small-business owner has to do is select the IT assets and the IT resources he or she needs or desires.

In a way, it can be a scary thing; after all, we are entrusting to store our valuable and confidential data to people at the ISP to whom the end user or the small-business owner has never even met or even heard of. However, it takes a giant leap of faith to have this kind of level and caliber of trust. However, despite this degree of uncertainty, the cloud-computing infrastructure does possess a number of quantifiable, key characteristics that help it define and separate itself from other IT infrastructure regimes.

These characteristics include

1. On-demand usage
2. Ubiquitous access
3. Multitenancy and resource pooling
4. Elasticity
5. Measured usage
6. Resiliency

On-Demand Usage

The first cloud computing characteristic, which is on-demand usage, means that if an end user or small-business owner provisions certain cloud computing services onto his or her account, this particular IT resource or IT asset will be made immediately for use after it has been provisioned. With the characteristic of on-demand usage, the end user can thus request future IT assets and resources on an automated basis, without any human intervention at the ISP whatsoever.

With this characteristic of an on-demand usage for the cloud-computing infrastructure, the end user or the small-business owner is literally free to turn on or off the IT assets and resources when he or she wants or needs and pays only for what is used.

Ubiquitous Access

The second characteristic of the cloud-computing infrastructure is known as ubiquitous access. This simply means that the IT resources and the IT assets are widely accessible from anywhere on earth and can be reached via any mobile device via any network protocol. This can best be illustrated by a small-business owner accessing his or her cloud-based IT assets and resources through the other side of the world, with his or her mobile device, via a wireless access protocol (WAP).

Resource Pooling

The third major characteristic of a cloud-computing infrastructure is multitenancy and resource pooling. The basic definition of *multitenancy* is "a characteristic of a software program that enables an instance of the program to serve different consumers (tenants) whereby each is isolated from the other" (*Cloud Computing: Concepts, Technology and Architecture*, Erl, T., 2013, Arcitura Education, p. 59). Take for example the Microsoft SQL Server.

There may be just one instance of it running on the physical server, but this particular instance allows to serve the different and multiple end users and small-business owners. This example demonstrates multitenancy and required the principles of virtualization to make it all happen. Although one instance of the SQL Server is on the physical server and shared with others, the end user or the small-business owner gets to feel that this is his or her very own software through his or her cloud-computing infrastructure control panel.

In fact, it is this multitenancy model that allows for the cloud-based IT assets and resources to be dynamically assigned over and over again, based upon the demands and the needs of the end user and the small-business owner. The multitenancy model, as a result, has given rise to another concept called resource pooling. This allows for "cloud providers to pool large-scale IT resources to serve multiple cloud consumers. Different physical and virtual IT resources are dynamically assigned and reassigned according to cloud consumer demand, typically followed by execution through statistical multiplexing" (*Cloud Computing: Concepts, Technology and Architecture*, Erl, T., 2013, Arcitura Education, p. 59).

These multitenancy and resource pooling concepts allow the end users and the small-business owners to use the same IT assets and resources while each end user or small-business owner remains unaware that the same IT assets and resources are being used by others as well.

Elasticity

The third major characteristic of a cloud-computing infrastructure is that of elasticity. When one thinks of elasticity, very often, the image of a rubber band is conjured up. This is perfectly analogous to what a cloud-computing infrastructure should be like. It should literally be able to flow smoothly and have the ability to scale to the needs and wants of the end user or the small-business owner.

For example, if an ISP expects to keep its customer base and attract newer technologies into its cloud-computing infrastructure, it must be flexible enough to keep up with the changing market conditions and demands that are placed upon it. An ISP that is not flexible and is rigid in its ways of technology will be sure to lose in terms of keeping its competitive advantage.

It should be noted that ISPs with vast resources for a cloud-computing infrastructure offer the most flexibility and elasticity. The elasticity of a cloud-computing infrastructure can also be best measured by how quickly an end user or a small-business owner can turn off or on his or her new cloud-based services.

MEASURED USAGE

The fourth major characteristic of a cloud-computing infrastructure is that of measured usage. As its name implies, this feature simply takes into account how much usage the end user or the small-business owner is taking out from the IT assets and resources that he or she has selected to use. Based upon that particular usage, the end user or small-business owner is appropriately charged for the cloud-computing infrastructure resources utilized.

Normally, this is a flat fee on a monthly basis, but if usage of the IT assets and resources go beyond the expected levels, the charges will be much higher. Conversely, if the usage of the IT assets and the IT resources falls below the expected usage level, the end user or the small-business owner still has to pay the flat monthly fee.

In this regard, a cloud-computing infrastructure possesses the *use it or lose it* feature, unless the specific services have been cancelled. However, the characteristic of measured usage does not necessarily mean how much to charge; it also refers to the overall usage of an IT asset or an IT resource being utilized and those not being used. That way, future demand can be predicted, and the cloud computing resources can be provisioned accordingly.

Resiliency

Finally, the fifth and final characteristic of a cloud-computing infrastructure is that of resiliency. This simply means that "IT resources can be pre-configured so that if one becomes deficient, processing is automatically handed over to another redundant implementation" (*Cloud Computing: Concepts, Technology and Architecture*, Erl, T., 2013, Arcitura Education, p. 61). In the scheme of cloud-computing infrastructure, it also means that there are redundant IT assets and resources available, so should one set fail, the end user or the small-business owner will have no downtime.

In other words, smooth and seamless operations are always present in a cloud-computing infrastructure.

Cloud-Computing Delivery Models

Now that we have reviewed some of the major characteristics of a cloud-computing infrastructure, it is time to understand how such a cloud-computing infrastructure can be brought into the hands of the end user or the small-business owner.

Although the cloud structure does involve being in a nebulous state, the delivery method is not so. There are many ways in which the cloud can be brought to the end user or the small-business owner, and there are three common types of platforms:

1. Infrastructure as a service (IaaS)
2. Platform as a service (PaaS)
3. Software as a service (SaaS)

Each of these platforms has its own unique role in delivering the cloud-computing infrastructure, and these platforms will be discussed at length in the following sections.

Infrastructure as a Service

The first cloud-computing infrastructure deployment model, which is the infrastructure as a service, also known as IaaS, represents the overall cloud-computing environment infrastructure. In other words, IT assets and resources "can be accessed and managed via cloud-based interfaces and tools" (*Cloud Computing: Concepts, Technology and Architecture*, Erl, T., 2013, Arcitura Education, p. 64).

As its name implies, the IaaS provides the framework or the foundation from which all of the IT assets and resources can be leveraged toward the small-business owner or the end user. An example of an IaaS is shown in the following figure. This infrastructure, in particular, includes the hardware, network connectivity, all of the software applications (which includes, for example, all of the VoIP applications, e-mail applications, database applications, software development applications, etc.) as well as other *raw* tools that comprise the IaaS Infrastructure.

It should be noted that most IaaS IT assets and resources are *virtualized* and bundled in such a package that they can be leveraged through the cloud to the end user or the small-business owner. Thus, these virtualized IT assets and resources can have the freedom of scalability, customization, and demand availability, which are all very crucial components of cloud-computing infrastructure.

By possessing an IaaS Infrastructure, the end user or the small-business owner can have total control and responsibility over their particular cloud-computing infrastructure. For example, once an end user or a small-business owner signs up for a cloud-computing infrastructure account, they are often given access to a control panel from which they can establish the settings and the permissions, and even install and uninstall particular cloud-computing resources.

Every ISP gives this tool to all of their customers. After all, this is the only direct way for end users or small-business owners to have access to all of the IT assets and resources to which they are subscribed. It should be noted that with the IaaS platform, end users or small-business owners assume full administrative control over their cloud-based IT assets and resources.

For example, when end users or small-business owners first sign up for their cloud-computing infrastructure account, the IT assets and the IT resources that they will be subscribing to are known as *fresh virtual instances*. Let us demonstrate with an example. Suppose that an ISP has literally hundreds of physical servers. These servers contain all of the IT software-based assets and resources.

Once the end users or small-business owners provision their own account, the hard drive from which the software-based IT assets and resources will be distributed will be divided into its own partition on the physical server. Thus, this will give the end users or small-business owners the look, the feel, and the total control over their own unique cloud computing server.

Conversely, if the end users or small-business owners need a gargantuan account of IT assets and resources, they can lease out an entire cloud-based server, which, in this scenario, is known as a *virtual server* in the IaaS platform.

Platform as a Service

The second deployment model for the cloud-computing infrastructure is known as PaaS. Specifically, it can be defined as "a predefined 'ready-to-use' environment typically comprised of already deployed and configured IT resources" (*Cloud Computing: Concepts, Technology and Architecture*, Erl, T., 2013, Arcitura Education, p. 65). The prime differentiation between the PaaS Platform and the IaaS Platform is that the latter consists of the raw cloud computing platform. An example of a PaaS can be seen in the following figure.

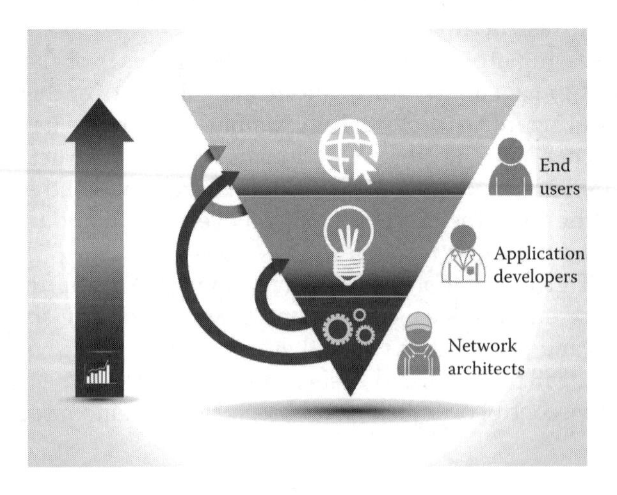

In other words, as reviewed previously, the IaaS contains the basic materials needed for the foundation of a cloud-computing infrastructure. Imagine the IaaS Platform that serves as the foundation for the cloud-computing infrastructure. The PaaS Platform fills up this foundation with the much needed IT assets and resources to fulfill the needs of the end user or the small-business owner.

As it can be described, the PaaS consists of a set of prepackaged IT products and IT tools to help support the business needs of the end user or the small-business owner. There are many reasons why a consumer should choose the PaaS, but the following are the most typical reasons:

1. Having the sense of scalability.
2. The client can use the literally ready-to-use environment and specifically modify it to his or her own needs and wants.
3. If the end user or small-business owner feels confident enough, he or she can even run his or her own services, through the use of the PaaS, to other cloud consumers.

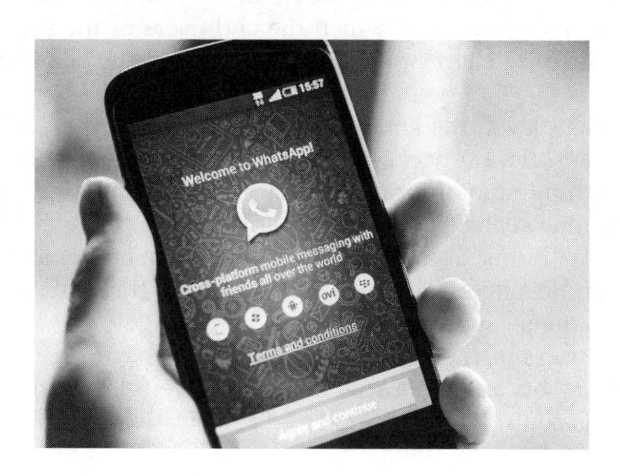

The PaaS permanently alleviates on the part of the end user or the small-business owner the need to fill up and have the overall responsibility of administering the IaaS. Rather, this option is up to the ISP to provide to its consumer base. Thus, as you can see, it is the PaaS that contains all of the needed IT assets and resources from which the end user or small-business owner can choose. To summarize thus far, the IaaS is a service that provides the foundation of a cloud-computing infrastructure, and the

PaaS makes the IT assets and resources available to the end user or the small-business owner either as a deployed cloud package or a single service. This is known as SaaS and will be discussed next.

Software as a Service

The next cloud-computing infrastructure to be looked at is SaaS. The IaaS provides the foundation for the cloud-computing infrastructure; the PaaS fills up this foundation with all of the IT assets and resources that would fulfill the needs of all types of end users and small-business owners. These end users and small-business owners will not need to use all of the IT assets and resources that are available in the PaaS; this is where the SaaS platform comes into play.

Specifically, the SaaS model in the cloud-computing infrastructure can be defined as "a software program positioned as a shared cloud service and made available as a 'product' or general utility that represents the typical profile of an SaaS offering" (*Cloud Computing: Concepts, Technology and Architecture,* Erl, T., 2013, Arcitura Education, p. 66). Thus, from the definition, the SaaS component of the cloud-computing infrastructure can be viewed as a *marketplace* in which the end users or the small-business owners can literally cherry-pick the IT assets and the IT resources they need or desire.

At the SaaS level, the small-business owners or the end users can pick all of the software packages or bundles they require to keep their business running smoothly and to maintain a competitive advantage. For example, one such SaaS offering that is popular is the hosted e-commerce store. With this, a small-business owner can attract many customers. However, despite all of these advantages of the SaaS, it does have serious limitations.

The end users or the small-business owners have very little control over their IT assets and resources that they have selected from the SaaS platform. This control is restricted to administrative control, after the IT assets and the IT resources have been selected and paid for. There is very little that can be done in administrating those IT assets and resources that reside in the SaaS.

In summary, we have looked at three major cloud delivery models:

1. *IaaS*: This lays the foundation or the nuts and bolts of the cloud-computing infrastructure.
2. *PaaS*: This model provides all of the IT assets and resources that an ISP can provide to its consumers. All of these resources and

assets include everything from the virtual server to the hard-drive partitions that provide the end users or the small-business owners with the look and feel of their own cloud-based server as well as the software applications, which range from content management systems to e-commerce platforms.
3. *SaaS*: As reviewed, this model provides the software applications (also known as the IT assets or IT resources) via an a la carte method to the end user or the small-business owner. Under this regime, the customer can pick and choose the software applications that are needed or desired, and within seconds, it can be provisioned at the click of the mouse after it has been paid for.

Cloud-Computing Deployment Models

Now that we have reviewed how a cloud-computing infrastructure can fit into a certain platform, we now turn our attention to how the cloud-computing infrastructure can be handed over to the small-business owner or the end user.

In other words, the cloud-computing infrastructure is now defined via three different infrastructure models, but now, it has to reach the power of the end users or the small-business owners via the network to their computer, laptop, netbook, or even smartphone. The cloud-computing infrastructure is brought to the end user or small-business owners via four different kinds of deployment models:

1. Public cloud
2. Community cloud
3. Private cloud
4. Hybrid cloud

We will now discuss each of these cloud-computing infrastructure deployment models in much greater detail.

Public Cloud

In terms of the cloud-computing infrastructure, the public cloud can be specifically defined as "a publicly accessible cloud environment owned by a third party cloud provider" (*Cloud Computing: Concepts, Technology and Architecture*, Erl, T., 2013, Arcitura Education, p. 73). An example of a public cloud can be seen in the following figure.

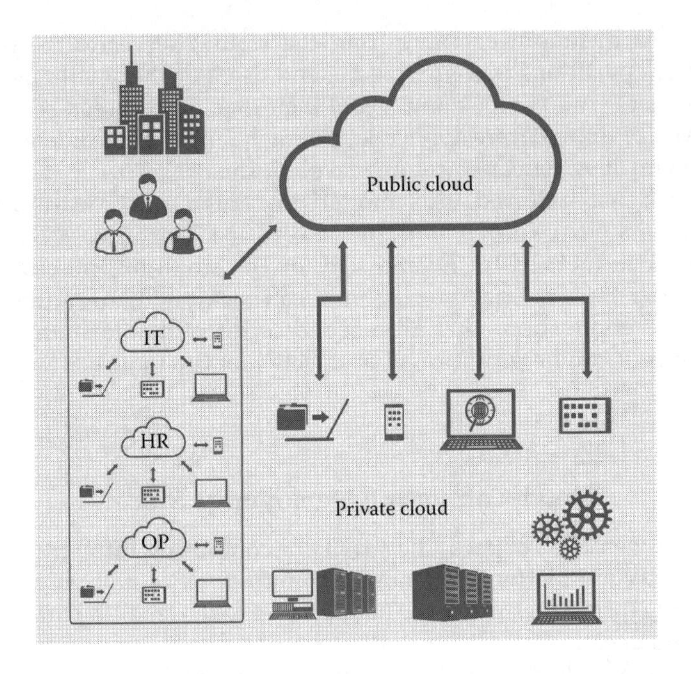

As the name implies, this type of cloud-computing infrastructure is available via any type of ISP. The IT assets and the IT resources, as described throughout this chapter, can be provisioned among all three delivery models just described (IaaS, PaaS, SaaS) and are also available to the end users and the small-business owners at almost very low cost or price; the IT assets and the IT resources are very often *commercialized* via other methods such as advertisements on the Internet, or in print, such as magazines and newspapers.

With the public cloud deployment model, the ISP has the primary responsibility of this kind of deployment, the acquisition and the procurement of the IT assets and resources that belong in the PaaS Platform, and its ongoing maintenance (such as licensing fees and any other necessary types of software and hardware upgrades).

It should be noted that all (about 98% or so) cloud-computing infrastructure offerings are offered to the small-business owner and the end user via this type of deployment model, so that all types of IT assets and resources can be made widely available and at low cost to all entities involved.

Community Cloud

The second type of deployment, which is that of the community cloud, is also very much related to that of the public cloud. However, access to this particular type of cloud-computing infrastructure is sheerly limited to a specific group of end users and small-business owners. It should be noted that, at this point, the community cloud could be jointly owned by the members of the cloud community, or even an ISP could own this community cloud.

The members of the community cloud typically share the responsibility for defining the rights and the responsibilities of the community cloud. However, it should be noted also that membership into the cloud community is not a status and that membership into it does guarantee access to the IT resources and the IT assets that reside in the cloud-computing infrastructure.

Private Cloud

The third type of cloud-computing deployment model is that of the private cloud. The private cloud-computing infrastructure is literally owned by a single entity. This type of cloud-computing infrastructure deployment model allows a small-business owner or any other kind of organization total control over the IT assets and the IT resources from within the different locations and departments at the place of business or organization.

Typically, under a private cloud-computing infrastructure deployment model, administration can take place internally, thus giving the private cloud owner even more sense of ownership and control. From within the private cloud-computing infrastructure deployment model, typically, the place of business or organization are both the consumer and the cloud owner.

To help the place of business define and differentiate these roles between the consumer and the cloud owner, the following guidelines are established:

1. A separate department from within the place of business or organization assumes the cloud provisioning duties to be performed. These groups of people become the private cloud owners.
2. All of the departments at the place of business requiring access to the IT assets and resources from within the private cloud-computing infrastructure become the cloud consumers.

Hybrid Cloud

The next type of cloud-computing infrastructure deployment model we examine is known as the *hybrid cloud*. As its name implies, the hybrid cloud consists of a combination of the various cloud deployment models just discussed, which include the community cloud, the public cloud, and the private cloud. The hybrid cloud does not have to consist of all these cloud types, but it must contain at least a combination of two types of cloud deployment models, and from that, any other cloud deployment model can work just as well, with the different deployments of IT assets and resources.

For example, a small-business owner can choose to protect his or her confidential and private data (such as financial information, customer data, etc.) from within the private cloud infrastructure. Now, if this small business grows in size over time to a multinational corporation, the business owners can then develop the private cloud into a hybrid cloud-computing infrastructure, put it into a community cloud, and create more if necessary.

As you can see, trying to implement a hybrid cloud deployment model can be very complex and much more of a challenge to administer and maintain because of the differing cloud environments that are presently available. Because of this, an alternative exists to the hybrid cloud—this is called the virtual private cloud.

Under this cloud regime, the end users or the small-business owners are given a self-contained cloud environment that is dedicated solely to their uses and are also given full administrative rights to do what the cloud end users or the small-business owners wish to do in their own particular virtual private cloud environment.

THE SECURITY THREATS POSED TO CLOUD COMPUTING

Now that we have covered in great detail and depth what the cloud computing delivery and deployment models are like, we turn our attention to the weaknesses that are posed to the cloud. More specifically, these can be categorized as *threat agents*. However, before we discuss what a threat agent is specifically for a cloud-computing infrastructure, it is first very important to review some very basic terms and concepts as it relates to cloud computing threat agents.

Specifically, these terms and concepts include the following:

1. Confidentiality
2. Integrity
3. Authenticity
4. Availability
5. Threat
6. Vulnerability
7. Risk
8. Security controls
9. Security mechanisms
10. Security policies

These above concepts and definitions form the basics for understanding what threat agents are to a cloud-computing infrastructure. The above definitions and concepts will be reviewed and examined in detail, but it is also equally important to review the threat agents posed to an overall cloud-computing infrastructure.

These threat agents include

1. Anonymous attacker
2. Malicious service agent
3. Trusted service attacker
4. Malicious insider

Apart from the above cloud-computing threat agents, there are also more specific security threats to a cloud-computing infrastructure:

1. Traffic eavesdropping
2. Malicious intermediary
3. Denial of service
4. Insufficient authorization
5. Virtualization attack
6. Overlapping trust boundaries
7. Flawed implementations
8. Security policy disparity
9. Contracts
10. Risk management

All of these concepts and terminologies will now be reviewed in much greater detail.

Confidentiality

With respect to the very basic terms and fundamentals, the first that is of importance is known as *confidentiality*. In simple terms, this means that the IT assets and the IT resources are only made available to authorized users. In this case, it means that the paying subscribers such as the end users and the small-business owners only have access to those IT assets and resources that they have paid for, especially when those resources and assets are in transit across the network medium, and after that, access remains very strict when it comes to remote logins.

Integrity

With regard to integrity, this means that the IT assets and the IT resources that reside at the ISP are not altered or changed in any way or form. However, more importantly, it is the end user's and the small-business owner's data and information that are being processed at the virtual servers located at the ISP that need strong assurance that this confidential and private information is not altered in any way or in any malicious form. In other words, "integrity can extend to how data is stored, processed, and retrieved by cloud services and cloud based IT resources" (*Cloud Computing: Concepts, Technology and Architecture*, Erl, T., 2013, Arcitura Education, p. 119).

Authenticity

With regard to authenticity, this proves that all of the IT assets and resources are legitimate and that all of the information and data processed at the virtual server at the ISP and all of the output that comes from it are genuine, and that all these have been provided to the end user or the small-business owner from a legitimate cloud-computing infrastructure provider.

Availability

In terms of availability, this means that the IT assets and the IT resources will always be accessible and be rendered usable whenever and wherever the end user or small-business owner needs to access them at a second's notice. This uptime is very often spelled out in the SLA to which the end user or the small-business owner must subscribe.

Threat

With regard to threats, it is a threat agent of sorts that can fatally challenge the defenses afforded to the cloud-computing infrastructure. It should be noted that the threat just defined if carried out with a malicious intent is known as a *cyber attack*.

Vulnerability

In terms of the next important security foundation, vulnerability is a weakness in the cloud-computing infrastructure that can be easily overcome by a cyber attack, as just previously described. A vulnerability in the cloud-computing infrastructure can be caused by a whole host of reasons, which can include both software and hardware misconfigurations, very weak security policies, administrative or user errors, hardware and software deficiencies, and an overall poor security architecture for the cloud-computing infrastructure.

Security Risk

In terms of security risk to the cloud-computing infrastructure, this represents the statistical probability of a loss occurring to the cloud-computing infrastructure. Cloud-computing risk is typically measured according to the threat level, or as a whole, when compared with the total number of possible or known threat agents. There are two known metrics used to determine the statistical probability of risk:

1. The statistical probability of a threat occurring to exploit the IT assets and the IT resources in the cloud-computing infrastructure
2. The chances of the expectations of a magnitude of a loss occurring from the IT resource or the IT asset within the cloud-computing infrastructure

Security Controls

With regard to security controls, there are countermeasures that are employed to protect the cloud-computing infrastructure and the IT assets and the IT resources that reside within it. The details on how to implement and use such security controls from within a cloud-computing infrastructure are often spelled out in the security policy.

Security Mechanisms

In terms of security mechanisms, there are countermeasures and protective mechanisms used to provide the general framework that protects the IT assets and the IT resources in the cloud-computing infrastructure.

Security Policies

Finally, the security policies associated with a cloud-computing infrastructure are the security rules and regulations that govern its safety. It is the cloud-computing security policy that furthers the types of rules and regulations that are needed to provide maximum security to the cloud-computing infrastructure.

Now that we have looked at some of the very important terms and concepts that underpin a cloud-computing infrastructure, as alluded to before, we now turn our attention over to the points of vulnerability. These are those specific entities who can carry out malicious activity against a cloud-computing infrastructure and provide great harm to the end user or the small-business owner.

The first type of these types of entities are known now specifically as the *threat agent*. As the term implies, this type of entity can carry out a direct attack against a cloud-computing infrastructure. What is even more dangerous about this type of attacker is that this entity can originate either internally within the place of business or organization or just even outside of it.

Anonymous Attacker

The first formal type of a threat agent is known as the *anonymous attacker*. This type of attacker is a malicious third party without any specific cloud resource-based permissions. This type of attacker exists as a software program and can launch specific attacks to the virtual servers via the public network. These types of attack entities have limited information or knowledge about the security defenses of the ISP, so the best way they can get into a cloud-computing infrastructure is by hacking into and stealing user accounts.

Malicious Service Agent

The second type of formal threat agent is known as the *malicious service agent*. This type of entity typically exists from within the cloud-computing infrastructure and has the capability to intercept and forward network traffic. Again, this type of threat agent usually exists as a software program.

Trusted Attacker

The third type of a formal threat agent is specifically known as the *trusted attacker*. As the name implies, the trusted attacker is already a consumer of a particular cloud-computing infrastructure, and therefore, they have advanced knowledge (to some degree) of its inner workings. This type of attacker usually launches their attacks from within a cloud-computing infrastructure, very often using their own login information from within a cloud-computing infrastructure, very often using their own login information and access credentials. The main threat of a trusted attacker is that given his or her knowledge of the cloud-computing infrastructure, he or she can attack just about every type of virtual servers from within the cloud-computing environment.

Malicious Insider

The final type of a threat agent is known as the *malicious insider*. Typically, these are contracted employees or ex-employees where login and access information into the particular cloud-computing infrastructure has not been deleted as of yet. Given this short window of timeframe until their access information is totally deleted, these types of attack entities can still log into their former cloud-computing infrastructures and cause great moments of harm and damage.

Traffic Eavesdropping

Now that we have seen what some of the major threat agents are to a cloud-computing infrastructure, we turn our attention to some of the major actual threats that are posed to a cloud-computing infrastructure. The first type of threat is known as *traffic eavesdropping*. This simply happens when the end user's or small-business owner's information and data that are still being transferred to their cloud-computing infrastructure are covertly intercepted by a third party (such as a malicious hacker) for the illegal purposes of misusing and abusing those particular information and data.

The pure purpose of this kind of attack is to breach the relationship between the end user/small-business owner and the ISP. The traffic eavesdropping security threat very often goes unnoticed for very long periods.

Malicious Intermediary

The second type of threat to a cloud-computing infrastructure is known as a *malicious intermediary*. In this type of attack scenario, the end user or the small-business owner's information and data are intercepted by a malicious third party and are intentionally altered, thereby compromising the integrity of those particular information and data, with the end result being totally spoofing the consumer of the integrity of that information and data.

Denial of Service

The third type of threat to a cloud-computing infrastructure is that of *denial of service*. The primary objective of this type of attack is to constantly overload and bombard both the physical and virtual servers up to the point they cannot operate anymore, and literally shut down, thus depriving the end user and small-business owner of their IT assets and resources.

Here is how a denial of service attack could conceivably work:

1. The physical servers and the virtual servers are totally bombarded with malformed data packets.
2. Subsequently, the network traffic is greatly reduced and responsiveness is totally choked off between the ISP and the end user/small-business owner connectivity to their IT assets and resources.
3. Next, the memory of both the physical and the virtual servers is totally bogged down by processing the malformed data packets and then shuts down, thus cutting off all types of access.

Insufficient Authorization

The fourth type of attack to a cloud-computing infrastructure is that of *insufficient authorization*. This is when an attacker is granted access by mistake by the ISP, thereby giving the attacker access to all of the physical and virtual servers and IT assets and resources. Another type of attack of this kind is known as *weak authentication*. This occurs when there is very low entropy or weak passwords that are being used by the end user or the small-business owner to access his or her specific IT assets and resources.

Virtualization Attack

The fifth type of attack to a cloud-computing infrastructure is that of the *virtualization attack.* With the nature of the cloud-computing infrastructure, the end user and the small-business owner have, to a certain degree, full administrative privileges. Because of this inherent risk, the end users or the small-business owners can become a cloud-computing infrastructure threat in and of themselves.

More specifically, the virtualization attack takes full advantage of the weaknesses of the confidentiality, integrity, and availability platforms of the virtual servers that reside within the cloud-computing infrastructure.

Overlapping Trust Boundaries

The fifth particular type of attack to a cloud-computing infrastructure is what is known as *overlapping trust boundaries.* One must remember that with a cloud-computing infrastructure, all of the IT assets and resources are shared among one another, even though the end users or the small-business owners get the look and feel as if it is their very own.

Given this shared resource nature, a small-business owner or end user who has an extensive IT background and a very strong malicious intent can easily take advantage of the vulnerability of shared resources, launch an attack, and totally corrupt the information and data, and even bring down the virtual servers of other end users and small-business owners. In other words, the goal here is to literally target the overlapping trust boundaries between the various cloud-computing infrastructures that exist at the ISP.

IMPORTANT MECHANISMS OF CLOUD COMPUTING

In this chapter, we have examined a number of concepts and theories that deal with a cloud-computing infrastructure. We have looked at everything from the various cloud models to the deployment mechanisms as well as some of the major threats and risks that are posed to a cloud-computing infrastructure. So far, in our examination of a cloud-computing infrastructure, we have looked at it in terms of a measured service, in other words, what it can do for the specific needs of the end user or the small-business owner.

However, the cloud itself has inner workings of its own that enable it to function the way it does. These components of the cloud-computing infrastructure can be referred to as specialized cloud mechanisms. In this part of the chapter, we review six major ones (there are many others, but these five are the most important), and they include the following:

1. The load balancer
2. Pay-per-use monitor
3. The audit monitor
4. The failover system
5. The hypervisor
6. Resource clustering

Load Balancers

With the load balancer mechanisms in the cloud-computing infrastructure, rather than place the strain on just one IT asset or just one IT resource, the workload is placed upon two or more IT resources to help increase the performance and capacity. The load balancers can perform the following workload distributions:

1. *Asymmetric distribution*: The larger workloads are given to those IT assets and resources that have more sophisticated processing power capabilities.
2. *Workload prioritization*: The cloud-computing infrastructure workloads are scheduled and distributed according to the priority level.
3. *Content-aware distribution*: The requests set forth by the end user and the small-business owner are sent over to the various IT asset and IT resources in the cloud-computing infrastructure.

It should be noted that the load balancer is preconfigured with performance rules for the optimization of the client-based IT assets and resources and to avoid overload in the cloud-computing infrastructure. These rules can exist as

1. A multilayered switch
2. A dedicated hardware appliance
3. Dedicated software
4. A service agent

It should be noted that the load balancer is usually located in the end user/small-business owner's IT assets and resources that are generating the workload and performing that actual workload.

Pay-per-Use Monitor

The second mechanism, the pay-per-use monitor, "measures cloud-based IT resource usage in accordance with predefined pricing parameters and generates usage logs for free calculations and billing purposes" (*Cloud Computing: Concepts, Technology and Architecture,* Erl, T., 2013, Arcitura Education, p. 184). This monitor can include

1. The data volume
2. The bandwidth consumption

It should be noted that the pay-per-use mechanism is monitored by the billing department of the ISP to transmit the proper fees to the end user or the small-business owner.

Audit Monitor

In terms of the third monitor, which is the audit monitor, this type of mechanism collects and analyzes various types of data and network traffic and activity within the cloud-computing infrastructure, which are primarily dictated by the regulations set forth in the SLA agreement.

Failover System

The fourth kind of cloud-computing infrastructure mechanism is that of the cloud-based failover system. With this type of system, it is used to increase the availability of the IT resources and the IT assets that are made available to the end user and the small-business owner. The basic premise is to provide large amounts of redundancy in case of a failover, so that all of the IT assets and the IT resources will always be available 24 × 7 × 365.

It is purely designed to rollover to another IT asset or IT resource in case one fails. In other words, the failover system provides redundancy so that all cloud-based assets and resources will always be available to the end user and the small-business owner. It should be noted that failover systems are used primarily in mission critical programs and applications. A redundant cloud-computing infrastructure mechanism can span thousands of miles across many countries and in many divisions of the ISP to provide the 24 × 7 × 365 redundancy for the IT assets and the IT resources. The failover or the redundancy systems for both the IT assets and the IT resources come in two basic types and configurations:

1. Active–active
2. Active–passive

With the active–active failover or redundant system, many instances of the same IT asset and IT resources act synchronously together. Then one type of failover occurs with one or more of the IT assets and resources, and then those failed IT assets and resources are removed from what is known as a load balancer scheduler. After the failed IT assets and resources are removed, then the remaining IT assets and resources still continue in the operational mode.

With the active–passive failover or redundant system, not all of the IT assets and resources remain active. Only a portion remains active. Thus, in this type of configuration, the inactive IT assets become available or active when any of the other IT assets and resources fail, and the processing workload is then redirected to those IT assets and resources that become active and functional.

It should be noted that in both types of failover or redundant systems, switching over to a new IT asset or IT resource is not done until at least a new IT asset or IT resource thoroughly dies out. However, there are cloud-computing infrastructure failover or redundant systems that can detect failover conditions before total failure occurs and literally shut down that IT asset or IT resource before it fails.

Hypervisor

The fifth very important type of cloud-computing infrastructure mechanism is that of the hypervisor. This is a very important tool in cloud computing, as the hypervisor is utilized to create or generate virtual servers as offset instances of the physical server system. It should be noted that the hypervisor can only be used on one type of physical server, and virtual images can be created of only that one particular server. Also, any IT asset or IT resources that are created from that hypervisor as a result of the virtual server generation are tied or associated only to that particular physical server.

Resource Clustering

The sixth cloud-computing infrastructure mechanism we will now review is that of resource clustering. Typically, all cloud-based IT assets and resources are set very far apart, geographically speaking. Despite this physical distance, these servers, in the virtual sense, can logically be combined into groups known as *resource clusters* to greatly improve the availability of IT assets and resources to the end user and small-business owner.

In this regard, it is the resource cluster mechanism that groups multiple IT assets and resources into a single cloud-computing infrastructure instance. This clustering effect increases cloud-computing capacity, load balancing, and the overall availability of the IT assets and the IT resources to the end user and the small-business owner.

There are three types of cloud-based cluster types:

1. Server cluster
2. Database cluster
3. Large data-set cluster

Server Cluster

With the server cluster scheme, both the physical and virtual servers are all clustered together in one large-scale effort to boost IT asset and resource performance and availability.

Database Cluster

In terms of database clustering in the cloud-computing infrastructure, this is designed and created to help improve the data availability that is owned by the end user or the small-business owner. The database cluster consists of a synchronization feature that helps to ensure the consistency of the data information that can be stored at various and different storage devices that are contained in the database cluster.

Large Data-Set Cluster

With a large data-set cluster, the data partitioning and the distribution are set forth so that the end user's and small-business owner's data sets can be efficiently partitioned without compromising the structure or the integrity of the data sets.

Cloud Computing Cost Metrics and Service Quality Mechanisms

Now that we have looked into what makes a cloud-computing infrastructure work from the inside as well as the outside, we turn our attention to another very important aspect of the cloud-computing infrastructure—which are the metrics that are associated with them. Why are metrics so important? Well, it is to give a sense of protection to both the end user and the small-business owner, knowing that there is something they can gauge or compare with other ISPs who are also offering the levels of IT assets and resources.

In other words, the end user and the small-business owner, given these metrics, can now quantitatively benchmark to see what will work best for them. In this chapter, we look at both cost metrics and service quality metrics. First, we look at cost metrics in detail, which include the following:

1. Network usage
2. Server usage
3. Cloud storage device usage
4. Other cost management considerations

Network Usage

In terms of network usage, it is defined as the "amount of data transferred over a network connection, network usage is typically calculated using separately measured inbound network usage traffic and outbound network usage traffic metrics in relation to cloud services or other IT resources" (*Cloud Computing: Concepts, Technology and Architecture*, Erl, T., 2013, Arcitura Education, p. 387).

As described in the definition, network usage consists of the following:

1. Inbound network usage metric
   a. It measures the amount of inbound network traffic.
   b. This metric is measured in bytes.
2. Outbound network usage metric
   a. This measures the amount of outbound network traffic.
   b. This metric is measured in bytes.
3. Intracloud WAN usage metric
   a. This metric measures the amount of network traffic of the various IT assets and resources that are located in geographic diverse segments.
   b. This metric is measured in bytes.

It should be noted that the network usage metric is ascertained by the following components:

1. Static IP address usage
2. Network load balancing
3. The virtual firewall

Server Usage

With regard to the server usage metric, it is defined as "measuring using common pay per use metrics in IaaS and PaaS environments that are quantified by the number of virtual servers and ready-made environments" (*Cloud Computing: Concepts, Technology and Architecture*, Erl, T., 2013, Arcitura Education, p. 389).

As it is defined, the server usage metric is divided into the following metrics:

1. The on-demand virtual machine instance allocation metric
2. The reserved virtual machine instance allocation metric

The first metric measures the usage fees in the short term, whereas in the latter, the usage fees are measured and calculated over the long term:

1. The on-demand virtual machine instance allocation metric
   a. It measures the availability of the Virtual Server instance.
   b. The measurement component is the virtual start date to virtual server stop date.
2. The reserved virtual machine instance allocation metric
   a. It measures the cost for securing a virtual server.
   b. The measurement component is the reservation start date to the date of expiration.

Cloud Storage Device Usage

In terms of cloud storage device usage, it can be defined as "the amount of space allocated within a predefined period, as measured by the on demand storage allocation metric. Similar to IaaS-based cost metrics, on-demand storage allocation fees are usually based upon short-term increments. Another common cost metric for cloud storage is the I/O data transferred, which measures the amount of transferred input and output data" (*Cloud Computing: Concepts, Technology and Architecture*, Erl, T., 2013, Arcitura Education, p. 390).

1. The on-demand storage space allocation metric
   a. It measures the time length as well as the size of the demand storage space.
   b. The measurement component is in terms of the date of the storage release and the reallocation to date of the storage allocation.
2. The I/O data-transferred metric
   a. It measures the amount of the data that are being transferred between the cloud-computing infrastructure.
   b. The measurement component is in bytes.

Other Cost Management Considerations

Along with the discussed usage metrics, there are also other cost management considerations that need to be described:

1. *Cloud service design/development*: This is when the pricing models and the cost templates are defined by the ISP.
2. *Cloud services deployment*: This is when the usage measurement is ascertained and put into place in the cloud-computing infrastructure.
3. *Cloud service contracting*: At this stage, the ISP determines the actual usage cost metrics.
4. *Cloud service offerings*: This is when the ISP formally publishes the actual costs for usage of the IT assets and the IT resources.
5. *Cloud service provisioning*: This is when the cloud service thresholds are established.
6. *Cloud service operations*: This is when the active usage of the cloud-computing infrastructure results in actual cost usage metric data.
7. *Cloud service provisioning*: This is when a cloud-computing infrastructure is decommissioned, and the cost data are archived and stored.

Now that we have looked at some very important costing metrics, let us turn our attention to the quality metrics that define a cloud-computing infrastructure. These quality metrics consist of the following components:

1. *Availability*: This refers to the service duration of the cloud-computing infrastructure.
2. *Reliability*: This refers to the minimum time between failures.
3. *Performance*: This refers to the response time and capacity of the cloud-computing infrastructure.
4. *Scalability*: This refers to the fluctuations in demand on the part of the end user or the small-business owner.
5. *Resiliency*: This refers to how quickly the cloud-computing infrastructure can be backed up and recovered in a time of disaster.
6. *Quantifiable*: This variable must be clearly established and absolute.
7. *Repeatability*: This variable is the ability to receive identical results.
8. *Comparability*: The units of measure (such as the amount of bytes) need to be standardized and made the same for all levels of comparison within the cloud-computing infrastructure.

9. *Easily attainable*: This should be a common-based form of measurement that can be most importantly used and understood by the end users and the small-business owners.

The cloud-computing infrastructure quality metrics are as follows:

1. Service availability metrics
2. Service reliability metrics
3. Service performance metrics
4. Service scalability metrics
5. Service resiliency metrics

Service Availability Metrics

In terms of the service availability metrics, it is composed of two submetrics:

1. *The availability rate metric*: This refers to the fact that an IT asset or IT resource can be expressed as a percentage of total uptime of the entire cloud-computing infrastructure.
   a. It is measured as a total percentage of service uptime.
   b. The measurement component is in total uptime.
2. *The outage duration metric*: This particular metric measures both maximum and average continuous outage service level targets.
   a. It is measured in terms of the duration of a single outage.
   b. The measurement component is the date/time of the start of the outage as well as the date/time of the end of the outage.

Service Reliability Metrics

The service reliability metrics can be defined as "the reliability that an IT resource can perform its intended function under predefined conditions without experiencing failure. Reliability focuses on how often the service performs as expected, which requires the service to remain in an operational and available state" (*Cloud Computing: Concepts, Technology and Architecture*, Erl, T., 2013, Arcitura Education, p. 407).

As can be seen from the definition, the service reliability metrics consist of primarily two metrics:

1. The meantime between failures
   a. It measures the expected time between failures.
   b. It is measured in terms of ISP operating times and the total number of failures.

2. The reliability rate metric can be defined as "the reliability rate that represents the percentage of successful service outcomes. This metric measures the effects of nonfatal errors and failures that occur during uptime periods" (*Cloud Computing: Concepts, Technology and Architecture,* Erl, T., 2013, Arcitura Education, p. 407).
   a. This measures the actual percentage of the successful outcomes under certain conditions.
   b. The measurement component is the total number of successful responses per the total number of requests.

Service Performance Metrics

The service performance metrics refers to "the ability on an IT resource to carry out its functions within expected parameters. This quality is measured using service capacity metrics, each of which focuses on a related measurable characteristic of IT resource capacity" (*Cloud Computing: Concepts, Technology and Architecture,* Erl, T., 2013, Arcitura Education, p. 407). This overall metric consists of the following specific metrics:

1. The network capacity metric
   a. The measurement component is the amount of total network capacity.
   b. It is measured in terms of throughput (bits per second).
2. The storage device component metric
   a. It is measured in terms of storage device capacity.
   b. It is measured in terms of gigabytes.
3. The server capacity metric
   a. The measurement component is that of the server capacity.
   b. It is measured in terms of CPU frequency in GHz and the RAM storage size in gigabytes.
4. The web application capacity metric
   a. The measurement component is the web application capacity.
   b. It is measured in terms of the rate of requests per minute.
5. The instance starting time metric
   a. The measurement component is the time required to start a new instance.
   b. It is measured in terms of the date/time of instance startup to the date/time of the start request.
6. The response time metric
   a. This measurement component is the time required to perform synchronous operations.

   b. It is measured by the total number of responses divided by the total number of requests.
7. The completion time metric
   a. The measurement component is the time required to start and finish an asynchronous task.
   b. It is measured by the date of requests and responses divided by the total number of requests.

Service Scalability Metrics

With regard to the service scalability metrics, they can be defined as "metrics which are related to IT resource elasticity capacity, which is related to the maximum capacity that an IT resource can achieve, as well as measurements of its ability to adapt to workload fluctuations" (*Cloud Computing: Concepts, Technology and Architecture,* Erl, T., 2013, Arcitura Education, p. 409). As it can be seen from the definition, it contains the following metrics:

1. The storage scalability
   a. The measurement components are the required changes to higher and more demanding workloads.
   b. It is measured in gigabytes.
2. The server scalability (horizontal)
   a. The measurement components are the server's capacity changes in the face of increased workloads.
   b. It is measured in terms of the number of virtual servers that are available.
3. The server scalability (vertical)
   a. The measurement components are the virtual server spikes.
   b. It is measured in terms of gigabytes.

Service Resiliency Metrics

In terms of the service resiliency metrics, it can be described as "the ability of an IT resource to recover from operational disturbances is often measured using service resiliency metrics. When resiliency is described within or in relation to SLA resiliency guarantees, it is often based on redundant implementations and resource replication over different physical locations, as well as various disaster recovery systems" (*Cloud Computing: Concepts, Technology and Architecture,* Erl, T., 2013, Arcitura Education, p. 411). This type of metric can occur at all phases of the cloud-computing infrastructure:

1. *The design phase*: This is how well prepared the cloud-computing infrastructure is to cope with challenges and disasters.
2. *The operational phase*: This refers to how well the differences are in the levels of service agreements during and after a serious downtime or even a service outage.
3. *Recovery phase*: This refers to the rate at which the IT assets and the IT resources can recover from downtime.

As it can be seen from the definition, there are two types of primary metrics here:

1. The mean time to switch over to metric
   a. The measurement component is the expected time to complete a virtual server switchover from a failure to another virtual server that is located in a totally different geographic region.
   b. It is measured in terms of switchover completion divided by the total number of failures.
2. The mean time system recovery metric
   a. The measurement component is the expected time to perform a complete recovery from a disaster.
   b. It is measured in terms of time of recovery divided by the total number of failures.

An Introduction to Biometrics in the Cloud

Now that we have an extensive overview of what the cloud-computing infrastructure actually entails, in this section, we will examine what biometrics in the cloud will look like. As previously mentioned at the beginning of this article, there are already a number of biometric applications that are using the cloud infrastructure in some shape or another.

However, these are not full blown, or complete implementations, but rather, they are just a microapplication (in other words, just using some aspect of the cloud). Also, the biometric technologies utilized are very limited in nature; primarily only voice recognition is being used.

The definition of biometrics in the cloud means that a full-blown biometrics infrastructure will be placed at the hands of the hosting provider. In addition, the type of application (and even the number of different kinds of applications) will be unlimited in scope, and this hosted biometrics infrastructure will be designed to fit the wide gamut of biometric applications that exist today and those that are under research and development.

This includes the biggest market drivers of physical access entry; time and attendance; single sign on; large-scale law enforcement (in particular, AFIS); the national ID card; the e-passport; as well as other types of verification and identification scenarios.

Also, this hosted biometrics infrastructure will be designed to support any type of biometric technology that is available today, as well as those that are also under research and development. This includes the physical biometrics (fingerprint recognition, face recognition, iris/retinal recognition, hand geometry recognition, and voice recognition) and the behavioral biometrics (keystroke recognition and signature recognition).

The concept of biometrics in the cloud will also be catered extensively to one very important component—the customer, especially that of the small-business owner. Anybody or any organization or any business, no matter how large or small they are, or wherever they are in the world, will be able to use cloud-based biometrics at any time that they are considering implementing a biometrics infrastructure at their place of business.

Therefore, biometrics in the cloud can be envisioned from two very different aspects: (1) the hosting provider and (2) the customer or the small-business owner. Both parties will have their own responsibilities to ensure that biometrics in the cloud does indeed become a reality.

First, let us look at this from the standpoint of the small-business owner. Imagine that a small-business owner wishes to implement a biometrics-based security system and is giving heavy consideration to the cloud-based option. How exactly will the small-business owner go about this process? We have to first assume that the small-business owner has no prior knowledge of biometrics whatsoever. Thus, it is safe to assume that they will have their own level of research of what needs to be done and most likely will have hired a biometrics consultant to explore all of the alternatives and possibilities of what can be done. This part will have included a major systems analysis and design of the small-business owner's existing security infrastructure and exactly what and where the role of biometrics can take place.

Next, the choices of what to do next will come down to two items: (1) whether or not to implement biometrics as part of the existing security infrastructure or to deploy it as a brand new implementation and (2) what type of biometric technology to utilize. The answer to the first question will depend primarily upon the results of the systems analysis and design just conducted. However, with the second question, with a cloud-based deployment, any biometrics technology should work.

However, there are also other important factors that need to be taken into account with this, such as ease of use, ease of deployment, ease of end user training, as well as cost affordability given that the small-business owner will most likely be on a stringent budget. With these variables taken into account, vein pattern recognition will be the choice of technology to use for biometrics in the cloud and is further discussed in A Detailed Example of How Biometrics in the Cloud Would Look Like.

Now that the small-business owner has chosen to utilize vein pattern recognition as his or her choice of technology for biometrics in the cloud, the next question is where to have the actual hardware devices installed at their respective place of business. Again, this is 100% dependent upon the application that will be utilized.

For example, if it is for physical access entry, the biometric devices will be obviously contained at the major access points at the business or organization; if it is for time and attendance, then the devices will need to be installed near where the employees work; or if it is for single sign on (which is for eliminating the use of passwords), then the biometric devices will have to be installed at each of the employee's workstations; and so on.

It is most likely that the small-business owner will have the actual biometrics vendor come out and install whatever number of vein pattern recognition devices is required. This is where the beauty of biometrics in the cloud comes into play. For example, under traditional deployments, installing and implementing a full-fledged biometric system can take a lot of time, effort, and exorbitant costs, especially on the part of the customer, if it is a large organization.

This is especially true if the biometric system is to be integrated into an existing or a legacy security system infrastructure at the place of business. However, with biometrics in the cloud, the biggest advantage is that a biometric system can be implemented either into an existing or legacy security infrastructure or into a brand new deployment with the greatest of ease.

For instance, all the small-business owner has to do is merely acquire the vein pattern recognition devices, install them, and ensure that the proper networking is configured correctly so that these devices can access the cloud-based biometrics infrastructure from wherever it is located in the world.

After all of the vein pattern recognition devices have been installed, and the proper networking has been ensured, the next major step for the small-business owners is to now activate their respective account with their biometrics in the cloud infrastructure plan with the hosting provider. It is envisioned that this will happen in a manner that is very similar when one chooses and activates a new plan with an Internet service provider, for such services as web hosting, e-mail, Internet telephony, backup storage, etc.

Of course, once the account is set up and activated, then the small-business owner can enter all of the names of the employees who will be using the system, the type(s) of application(s) for which biometrics in the cloud will be used, and all the other relevant services that come with a traditional biometrics deployment. This includes such things as verification/identification services, template reporting and transaction services, administration services, the relevant software applications, and many more.

It is anticipated that from when the time the devices are actually installed and activated, it will take just a matter of minutes until a full blown, cloud-based biometrics application will be made available to the small-business owner. Again, this is one of the major benefits of biometrics in the cloud. For example, because of the hosted nature, a complete biometrics implementation and application can be set up in just a matter of minutes, as opposed to the days or even weeks with the traditional methodologies of biometric system deployment and implementation.

However, an important point must be noted here. It is assumed that the small-business owner will have established a relationship with a hosting provider and have the account set up before the actual devices are installed. In the beginning, there may not be a lot of hosting providers offering cloud-based biometric services; thus, the small-business owners may be very limited in their choice of where their biometrics infrastructure will be hosted. However, it is anticipated that over time, and as the demand for biometrics in the cloud increases, the number of hosting providers will proliferate much like that on the level of ISPs.

Now that we have examined the concept of biometrics in the cloud from the standpoint of the end user, most notably, the small-business owner, it is equally important, if not more, to look at all of this from the perspective of the hosting provider. From this standpoint, one of the first questions to be asked will be is whether the existing ISPs will take on this new role of hosting a biometrics infrastructure or will brand new ISPs have to be created and dedicated to just doing this?

Without conducting an extensive cost–benefit analysis, it will be very difficult to answer this question in detail. However, at first glance, it would make much more sense for the existing ISPs to take on this added role, at least initially. After all, the necessary hardware is already in existence, and at the beginning stages, it is anticipated that only software modifications and enhancements will have to be done to support a cloud-based biometrics infrastructure.

Another point that needs to be taken into consideration is that existing ISPs are not simply going to jump on board to support biometrics applications. After all, this is a huge change in the way they currently do things, and this will incur changes in the existing processes that will no doubt require deep levels of testing before anything goes live to the end user.

In this regard, it will be very crucial for the biometrics vendors as well as the industry as a whole to *make their case* to these ISPs and to work with them very closely to ensure that the overall biometrics in the cloud infrastructure will be launched successfully and smoothly.

In other words, strong partnerships and alliances have to be formed between the ISPs and the biometrics vendors. Once a handful of leading ISPs have been identified, the next step will be for these ISPs to take now on the added role of hosting a cloud-based biometrics infrastructure. Thus, it will take crystal clear communications and very clear planning to ensure that this transitory step occurs as smoothly as possible.

Also, herein lies another strategic advantage of biometrics in the cloud. For example, as this infrastructure starts to roll out with the ISPs, except for the software modifications and some minor hardware changes, this concept will already assume the existing cloud infrastructure described earlier in this section. This infrastructure includes PaaS, SaaS, and IaaS.

In further analyzing these three components that will support a biometrics in the cloud infrastructure, it is crucial to take a top-down approach. Thus, in this regard, it is the IaaS that can be viewed as the overarching layer in this model. In fact, one can consider the IaaS to be the underlying architecture for which the next two components of SaaS and PaaS will be built upon. The IaaS consists of the actual servers, storage (including the databases), networks, and other major computing resources. Thus, as you can see, it is the IaaS that is the heart of biometrics in the cloud.

However, before we delve further into how the subcomponents of the IaaS will support a cloud-based biometrics infrastructure, one very important point should be made here. Although the actual servers and other relevant hardware are contained at the ISP, the actual raw storage units are further partitioned into smaller subsections from within the hard drive of the server.

These smaller subsections are referred to as *virtual servers.* In other words, these virtual servers contain all of the software resources that the actual, physical server has, thus giving the small-business owner the look and feel of having his or her very own server from within the control panel of his or her hosting account. As a result of this virtual server, biometrics in the cloud thus possesses one of its strongest advantages, that is, low, predictable, and monthly fixed costs, which makes this whole cloud infrastructure very affordable to the small-business owner.

The IaaS component would consist of the following features, which in turn, would provide the underlying architecture for biometrics in the cloud.

Virtual Servers

This is the main point of *contact* if you will, where the small-business owners will have total control over their biometrics in the cloud infrastructure. At this level, the small-business owner will have a control panel (much like a web/e-mail hosting control panel) from which he or she will be able to oversee all transactions and have total administrative control over his or her cloud-based application. The underlying premise here is that the customer can essentially do whatever he or she needs to or desire to for his or her cloud-based biometrics infrastructure without intervention. However, the customer will not be able to do any kind or type of software upgrades or modifications to his or her virtual server, as it will be up to the hosting provider to do all of this. The key component of the virtual server is the operating system, and this will support both open- and closed-source platforms (such as Linux and Windows, respectively), as described in the following.

On another note, should the small-business owner require technical assistance of any kind, support will come from two sources: (1) the hosting provider itself (for technical assistance with the underlying hardware and software) and (2) the biometrics vendor(s) who are working with the ISP (in this case, the biometrics vendor(s) will provide technical assistance in terms of the biometrics perspective in these cloud applications).

Storage

This will be the place where all of the biometric templates will be stored, as well as the associated databases that will process the various transactions of these templates. In this regard, there are several key points to be made here. First, the small-business owner will have full development freedom in how he or she wishes to design and create his or her required databases. Meaning, he or she will have a choice over the technology platform from which upon the respective databases will be built.

The small-business owner will have his or her choice if he or she wants to utilize either a closed- or open-source database model. The former refers to UNIX-based architectures such as MySQL and POSTGREsql, and the latter refers to Windows-based architectures as SQL Server and other Oracle-based architectures. Of course, closed-source databases will mean an extra expense for the ISP in terms of software licenses and upgrades. Thus, the small-business owner will have to pay the respective fees to the ISP to contain his or her costs.

In these databases, the biometric templates to be stored will include the enrollment and verification templates of each employee that the small-business owner has. Of course, depending upon the development that is done, robust statistical and reporting features will be made available to the small-business owner as well, so that he or she can analyze the effects and impacts of his or her cloud-based biometrics infrastructure.

Also, herein lies the other great advantages of biometrics in the cloud: scalability. For example, in traditional biometric deployments, the databases have to be literally redesigned and rebuilt to cope with the demand for increased scalability.

However, with the cloud, the biometrics database(s) can be scaled to fit any array of biometrics applications within just a matter of minutes for a wide array of applications, ranging from the simplest 1:1 verification scenarios to the most complex 1:N law enforcement identification applications.

Networks

This is the third major component of the IaaS. Under this, a biometrics in the cloud infrastructure will have network connectivity to the vein pattern recognition devices at the physical location of the small-business owner, and vice versa. However, for each side to recognize each other (the hosting provider and the vein pattern recognition devices), each cloud-based biometrics application will have its own unique Internet Protocol (IP) address so that the small-business owner and his or her respective cloud-based infrastructure can be uniquely identified in the vast expanse of the Internet.

Also, given the shared choice of open- and closed-source software platforms as described above, the appropriate networking protocols will also have to be offered by the ISP to the small-business owner. However, it should also be kept in mind that the networking component will be one of the most critical features of biometrics in the cloud. This is because the 2-s-or-less verification times that have been realized in traditional biometrics applications will also have to be realized in cloud-based applications. This is especially important in the case when massive, law enforcement identification applications are applied to the cloud as well. Gargantuan searches through millions of biometric template records will have to come down to just a matter of minutes.

Thus, as one can see, network and bandwidth optimization will be of upmost importance, as well as the processing power of the virtual servers.

In addition, there is the question of just how safe the biometric templates are when they are in transit across the network.

For example, do they need to be encrypted? To allay these concerns and fears for the small-business owner, VPNs will also be offered. With this, the data packet that stores the biometric template will be further encapsulated by another data packet, thus providing greater assurances of security.

The second major component of the biometrics in the cloud infrastructure is the SaaS. In the cloud world, generally speaking, SaaS is fast becoming a popular choice. In biometrics, it is becoming very popular as well. However, with biometrics in the cloud, the SaaS component will come from two different angles: (1) software, which is already created and can be purchased on demand by the small-business owner; and (2) a software platform in which full, customized development will take place.

Keep in mind that, at this point, it is the SaaS component that the small-business owner will come into contact the most in terms of his or her hosted biometrics infrastructure. Therefore, just like the networking aspect as described previously, the SaaS also has to work very smoothly with almost 100% uptime (all of the relevant uptimes for biometrics in the cloud will be spelled out in the SLA agreement).

With respect to the first software option, it is envisioned that the biometrics vendors will create and develop software applications for all of the biometric technologies that are available today, both in terms of physical biometrics and behavioral biometrics. The applications developed will fit all of the major market categories including physical access entry, time and attendance, single sign on, etc.

Of course, these applications will not be developed for free. The biometrics vendors will sell these to the ISPs and provide the relevant support that would be needed. In turn, the ISPs would make these software applications available on demand to the small-business owner for a fixed, monthly fee. It could also be the case that these software applications could also involve some certain degree of customization to better fit the needs of the small-business owner.

With regard to the customized software platform, this will give the small-business owner the opportunity to create and develop his or her own biometrics applications. It should be noted that the ISP will not provide the development; rather, the small-business owner will have to hire his or her own developers. In this aspect, both open- and closed-source platforms will be provided.

With the open-source model, probably the most popular development tool will be that of PHP, Perl, etc., and with the closed-source model, the most widely used tool will be that of ASP.Net, as well as other software and content packages such as Sharepoint. At this stage, it is envisioned that the biometrics software applications created will be web-based.

After all, it is the Internet that will be what the small-business owner used to access his or her cloud-based biometrics infrastructure. Thus, web server software such as Apache (for open source) and Internet Information Services (IIS, for closed source) will also be offered.

Also, with any software applications comes the backend—namely, the databases. Thus, the IaaS and SaaS levels will have to work closely with one another. With regard to the SaaS component, there will be one thing that will be very important, and that is having a set of common standards and best practices for software development of applications for biometrics in the cloud. Currently, in the biometrics world, there is no central authority that oversees the many individual and disjointed biometrics standards and best practices that currently exist globally.

Also, at the present moment, strictly speaking, the cloud, in general, is a huge nebulous of virtual servers, TCP/IP addresses, networks, and literally hundreds and thousands of applications, all coming together to very much resemble a jagged puzzle. Therefore, as biometrics in the cloud does over time proliferate in terms of adoption on a global scale, the need to have a central authority to oversee and enforce a list of best practices and standards will become ever so important, with some substantial advantages to be gained.

For example, the small-business owner can look up to this list as a benchmark and guide as to how his or her biometrics in the cloud software applications should be developed. However, the biggest benefit will be that this open dialogue (as it is envisioned) will foster a sense of collaboration and support throughout the biometrics community worldwide.

A real-world example of how this would look is the BioAPI Consortium. Currently, this is a grouping of some 100+ biometrics vendors worldwide who have come together to create and support a standard Application Programming Interface (API), which will support all types of biometrics technologies available today, in terms of software development.

The third major component of the biometrics in the cloud is PaaS. The PaaS itself is not a unique structure like the SaaS or the IaaS because it is still a vehicle for software development. However, the main difference is the magnitude upon which these biometrics software applications can be built.

For example, the SaaS will support on demand software applications and customized software development, as described previously. With the latter, the intention is to create those types of biometrics software applications that can support the needs of the small- to medium-sized businesses and organizations. The best examples of this are the one-to-one verification software applications.

However, where would the much larger scale 1:1 verification and the 1:N biometrics software applications reside? It would be in the PaaS. In other words, the PaaS can be viewed as an extension of the SaaS component from which much more complex software applications can be created and supported.

Just like the SaaS, the PaaS will also support both closed- and open-source software development models. However, the main difference between the two is that the PaaS will be much more exhaustive in terms of development resources to support the much more complex biometrics software applications.

There are a number of large-scale, biometrics applications that would not just be a candidate as a cloud-based application but would reside very well within the PaaS:

- *The national ID card*: There are various nations around the world that have started to use this, and it is a method for citizens of a particular country to prove their citizenship. However, the national ID card has become much more than its original, intended purpose. It has become a document of sorts for which it can be used for e-voting, filing medical insurance claims, claiming government-based entitlements and benefits, etc.

  However, it is not a just a card—it is a smart card that can contain many biometric templates, ranging from fingerprint recognition templates all the way to facial recognition templates. As a result, this biometrics aspect of the national ID card could fit very well into the PaaS structure because large-scale software applications would have to be developed to support the millions upon millions of the people in a nation who use it.
- *The e-passport*: This is now taking the role of the traditional paper passport for the identification of a foreign traveler. Just like the national ID card, the e-passport is really nothing but a smart card with biometric templates stored in it (and the storage space is quite high). To use the e-passport, the passenger merely has to point their e-passport to a reader (via a wireless communications

channel), and if 100% verified, the traveler is allowed to enter the country of their final destination. Of course, the traveler has to submit their biometric templates at their country of origin. Because of this biometric aspect to the e-passport, it would be a perfect application for a biometrics in the cloud application, as well as for the PaaS, just given the sheer number of biometric transactions that have to occur each second in every airport around the world.

- *Law enforcement*: In the United States, it is the AFIS database that has supported law enforcement needs at both the state and federal levels. AFIS stands for Automated Fingerprint Identification System and is a gargantuan fingerprint template database of known suspects, criminals, terrorists, etc. There are current efforts underway to update this legacy database. At the same time, the technology associated with law enforcement has advanced greatly, and it is at the point now where wireless devices are used by law enforcement officials to collect the fingerprints of an apprehended individual in real time. This is then transmitted to the AFIS database to confirm the identity of the individual in question. However, there are time delays and processing wait times for these biometric transactions to take place, and the subsequent results presented to the law enforcement officer. Because of this, the AFIS database, if migrated to a biometrics in the cloud infrastructure, would drastically reduce these transaction times from minutes to seconds. Also, the massive software applications that are needed at the present time to support the AFIS database will fit very well into the PaaS component of cloud-based biometrics applications.

It should be noted at this point that these large-scale biometrics applications, as well as others, cannot suffer any downtime whatsoever, both from a network connectivity and infrastructure standpoint. As a result, redundancy is very important here, and in the case of any outage, all biometrics in the cloud resources would simply be rolled over to the backup IaaS, Paas, and SaaS. This is one of the other great advantages of biometrics in the cloud. Given its pooled and shared resources nature, redundancy is much more possible, as opposed to if these infrastructures were physically housed within the entities themselves.

Now that the major concepts of the biometrics in the cloud infrastructure have been reviewed, the next major question to be looked at is what kind of cloud model should be offered. Four different types of cloud models were discussed earlier in this section. Thus, given that the premise of the biometrics in the cloud infrastructure is to be made available to any entity, any organization, at anytime, anywhere in the world, and on demand, the public cloud model will work best.

The only criterion that will be used in this cloud model is if the business entity or organization has the financial means to purchase the desired or needed cloud-based biometrics applications. Today, the primary cloud infrastructure offered by the various ISPs is that of the public model. The only discrepancy is that of pricing. Some ISPs offer free services, some offer services at exorbitant prices, and some are at the middle of the road of the pricing scheme.

Therefore, when choosing the ISP for their cloud-based biometrics infrastructure, the small-business owner should remember the old adage *you get what you pay for*. However, also, it should be kept in mind that the ISPs should not be charging exorbitant prices either, and this will have to be carefully monitored. One of the ultimate goals of biometrics in the cloud is to have the ability to offer small-business owners affordable levels of service (as well as enterprise grade), as opposed to the high cost of traditional biometric deployments.

A DETAILED EXAMPLE OF HOW BIOMETRICS IN THE CLOUD WOULD LOOK LIKE

This section will now put both together into a real-world example of what biometrics in the cloud will look like. Again, the two major components throughout this entire process are (1) the small-business owner (or really any other customer for that matter) and (2) the hosting provider in which the biometrics in the cloud infrastructure would reside. First, let us begin this illustration from the standpoint of the small-business owner.

It is safely assumed that the small-business owner has some sort of security need—thus, he or she is exploring a cloud-based biometrics solution for his or her needs. After conducting a comprehensive requirements analysis (of course, with the help of a biometrics consultant), the small-business owner decides to implement a cloud-based biometrics solution.

Further, it has been decided that it will be a brand new security solution. Meaning, it will not be part of an existing security infrastructure; rather, this new cloud-based solution will be a primary means of security, not depending upon other legacy security systems. This new solution will be used for physical access entry, in which the small-business owner will allow employee entry into the place of business strictly based upon his or her ability to be successfully verified by the biometrics system.

Also, it is assumed in this scenario that the small-business owner has cost considerations; the small-business owner would like to have a biometrics system that will be easily and readily accepted by his or her employees; the employee who is training to learn how to use the system will have a very low learning curve; and the small-business owner has a system that can be very easily and quickly installed, imposing no further impediments to existing business processes.

With all of this in mind, the small-business owner ultimately chooses vein pattern recognition, and he or she has also picked out the supplier who will come to install the hardware, as well as the requisite networking that will be needed for the vein pattern recognition hardware to link up with the cloud-based biometrics infrastructure located at the hosting provider.

For simplicity of this illustration, assume that the small-business owner decides to install just one vein pattern recognition device at the main point of entry at the place of their business, and that the number of employees is small—say, about 20 employees. All that is needed is to have the actual vein pattern recognition device installed, the networking completed, and the employees trained in how to use the system.

However, there is one more step that needs to be completed after all of this, and that is the setup of the biometrics in the cloud hosting account and the infrastructure configuration. This part will be examined in more detail after the second piece is now examined—the hosting provider.

From the standpoint of the hosting provider, the biometrics in the cloud infrastructure will actually be quite simple when compared with the many other complex applications that can be hosted. Since the business is small, the application needs will be small as well. For example, in this scenario, all that will be needed is 1:1 verification.

Thus, as a result, the only cloud components that will be needed are the IaaS and the SaaS. However, it should be noted that as the security needs of the small-business owner grow and expand over time, more components and services from the cloud-based biometrics infrastructure can be added on demand.

For instance, should the small-business owner drastically expand their business into a much larger one, to include applications that involve 1:N identification scenarios, the subcomponents of the PaaS can be implemented within just a matter of minutes. Likewise, the opposite is also true. Should this small-business owner decide to cut back on the components and services for which he or she originally signed up, these can be removed on demand as well.

Looking at the standpoint from the IaaS, the small-business owner will need the following:

The virtual server: This will be the heart of the biometrics in the cloud infrastructure. It is at this point that the small-business owner will have total control over his or her entire infrastructure.

Storage: This part will consist of the database upon which the biometrics templates will be stored. Since the needs are quite simple, only one database will be required. The small-business owner, of course, will have his or her choice of technology from which to build his or her database, but it is safe to assume at this point that the initial design will be kept quite simple. Once the virtual server is configured, and the employees have started to enroll into the vein pattern recognition reader, the first types of data to be stored in this database will be the enrollment templates.

Networking: The importance of this subcomponent cannot be emphasized enough. It is essential that the networking is working at peak performance everyday and every second. For this example, the standard networking protocol of TCP/IP will be used (this will be the default protocol used, unless the small-business owner specifies otherwise) to communicate between the vein pattern recognition device and the cloud-based biometrics infrastructure.

Looking in terms of the SaaS component, this is what will be needed:

Software: This will be another, subsequently important piece for the small-business owner once everything is all set up. It should be noted at this point that the IaaS will have to be configured first before the SaaS subcomponent can be established. Given such simple needs in this example, the question often arises if any software application is needed at all. However, let us assume at this point that the small-business owner would like to have a software application that is designed for vein pattern recognition and for 1:1 verification scenarios. By having this, the small-business owner can see first hand how well his or her biometric system is actually working. For example, he or she can require robust statistical reporting features, which can help him or her optimize the system even more. Also, because of these simple needs, the small-business owner decides to go with a software application that is already predeveloped; therefore, there is no need for any customized or specialized software development.

Now that both of the sides have completed their side of the setup (remember, the hosting provider will always have their biometrics in the cloud services ready literally, 24 × 7 × 365), the two now have to be linked to each other.

This linkage will be completed through the control panel that the hosting provider has given to the small-business owner. Through this control panel, during initial configuration of the virtual server, the small-business owner will be guided through a series of steps that will lead them to the ultimate setup of their biometrics in the cloud infrastructure.

The following are some of the major steps the small-business owner will have to complete:

1. *The type of application to be created*: The small-business owner will select physical access entry, time and attendance, single on, etc.
2. *The type of biometrics devices installed*: The small-business owner will select from a list of choices of what types of biometrics he or she has installed to support his or her cloud-based biometrics infrastructure as well as the total number that is installed.
3. *The type of matching services needed*: The small-business owner will select if he or she needs 1:1 verification or 1:N identification matching services for his or her application.
4. *The number of employees or people who will be enrolled in the system*: The small-business owner will have to enter how many people will be enrolled thus far.
5. *If any software applications will be needed*: The small-business owner will be asked to select if he or she wants to have a predeveloped software application or develop one on his or her own. Should he or she choose the former, it will be ready to go immediately after the configuration has been completed.
6. *If any databases will be needed*: The small-business owner will either select yes or no; if he or she selects yes, he or she will then be presented with a list of database technologies he or she chooses from to create his or her database(s).
7. *Agreement to terms of service and contract; credit card information entered*: Before the cloud-based biometrics infrastructure, the small-business owner has to agree to abide by the contract, and a credit card will be the primary means of payment (of course, he or she will have various payment options as well, such as monthly, yearly, etc.).

Once all of the above has been completed and the small-business owner finally clicks on the *submit* button, the biometrics in the cloud infrastructure will now be set up and ready to go. However, a number of key points have to be made here. First, the above steps are just a sampling; it is quite conceivable that many other options could be presented to the small-business owner during this process.

Second, the hosting provider will require the small-business owner to submit a series of test biometric templates after initial setup to ensure that all is running smoothly and without any errors before going live. Third, this biometrics in the cloud control panel will also have another area from which the small-business owner can select the other services and add-ons, as needed (such as database setup/configuration; technical support; running certain types of tests on prototype applications; software development/configuration, etc.).

Fourth, many ISPs offer their customers certain web hosting packages that fit their needs and budgetary constraints. However, with the biometrics in the cloud infrastructure, this will be different. Meaning, once the small-business owner has activated his or her account and completed the initial virtual server configuration, he or she will have access to all of the services and options, right from the beginning. In other words, the biometrics in the cloud infrastructure will be all inclusive; there will be no different packages to choose from.

The small-business owner can always upgrade or downgrade his or her services and options on demand and will only be charged for the services he or she has rendered. Fifth, with this all-inclusive nature, the small-business owner can host many other different biometrics applications from within this one infrastructure.

A REVIEW OF THE ADVANTAGES AND THE DISADVANTAGES OF BIOMETRICS IN THE CLOUD

At this point, now that the major concepts of biometrics in the cloud have been examined and an actual example provided, it is important to review the major advantages and disadvantages of biometrics in the cloud.

The advantages include the following:

1. *An instant setup of a biometrics infrastructure*: Within literally minutes, this entire system can be set up all at the click of a mouse. The only exception to this is the time delay that will be experienced when the biometric devices are installed at the place of business or organization.

2. *It is on demand*: The biometrics services and other components can be added on or cancelled instantaneously.
3. *It is affordable*: This is especially the case for the small- to medium-sized businesses. Because of the cloud-based infrastructure, any costs will be available at a fixed and predictable monthly price; this is unlike traditional biometric systems where the costs can greatly escalate over a short period.
4. *It is highly scalable*: With this in mind, a biometrics application can be cut back or expanded greatly in just a matter of seconds; with the traditional biometric deployments, upgrades or downgrades can mean an entire revamp of the whole system, thus making it virtually cost prohibitive to do.
5. *An entire community of technical support and collaboration*: Given the open-source nature of biometrics in the cloud and that a set of best practices and standards will have to be adopted, customers from all walks of businesses and enterprise can readily share their experiences and expertise with others who are contemplating launching a biometrics in the cloud infrastructure.
6. *The business owner has no responsibility for hardware or software*: Except for the cost of the biometric device hardware, the small-business owner assumes no more responsibility for this; this is the ultimate responsibility of the hosting provider for upgrading the hardware and software for the biometrics in the cloud infrastructure, including software licensing fees as well as maintenance costs.
7. *Interface with legacy systems*: A cloud-based biometrics infrastructure can fit quite easily and readily into an existing security infrastructure should the small-business owner choose to have this option (in other words, a cloud-based biometrics application would provide for a superb multimodal security solution).
8. *An existing infrastructure in place*: For those ISPs that will be offering a biometrics in the cloud infrastructure, the transition should prove to be a relatively smooth one; this is so because all of the existing hardware and software that the ISP has will be used to create a biometrics in the cloud infrastructure.
9. *Redundancy is very easy*: Given the pooled and shared resource nature of the biometrics in the cloud infrastructure, redundancy is thus very easy and very cost effective; in traditional biometric deployments, redundancy very often means extra servers to store the biometric templates and other processes, which means a much greater expense.

10. *An enterprise grade-level biometrics system is available to all*: Traditionally, those biometric systems with the best performance and peak sophistication were only available to those businesses who could afford it—namely, the Fortune 500 companies. Now, with biometrics in the cloud, an enterprise-grade biometrics system can be made available to all, and not just a select few.
11. *New cloud-based services could emerge*: For example, forensics makes heavy use of biometric evidence that is left behind at a crime scene and is collected later by experts. The most common type of this evidence is what is known as *latent fingerprints*. Therefore, an offshoot of biometrics in the cloud could very well be what is known as *forensics as a service*.

The disadvantages include the following:

1. *Time*: This is probably the biggest disadvantage for biometrics in the cloud. For instance, biometrics has a very slow adoption and acceptance rate, especially in the United States and in Europe.
2. *A long lag time for the ISPs*: The ISPs also will have a very long time to *warm up* to the idea of offering a cloud-based biometrics infrastructure. For example, many ISPs today just offer hosted office-based solutions for their customers, which is where the demand is right now. It will take a lot of work and convincing on the part of the biometrics industry to prove to the ISPs why they should offer a biometrics in the cloud infrastructure for their customers, as well as the positive return on investment they should experience over the course of time.
3. *Changes in ISP business processes*: Although it is expected that the cost of any extra hardware or software imposed upon the ISPs should be minimal, it is also expected that the ISPs will be very hesitant from the outset about implementing biometrics in the cloud; after all, this means a drastic shift from the current business processes because it is just human nature in general to be very hesitant about positive change in general.
4. *Changes in biometric business processes*: At present, every biometrics vendor has his or her own production processes, research and development strategies, and corporate objectives. For instance, some vendors offer different products and services. However, when biometrics in the cloud indeed become a reality, the entire biometrics industry as we know it will have to shift their strategies to an almost services-based environment.

5. *A cloud-based biometrics infrastructure needs to be scalable to demand*: When biometrics in the cloud becomes a reality, it is very important that the ISPs not devote entire resources just yet to cloud-based biometrics. Meaning, there should not be idle cloud-based resources expecting to be consumed when they may never be. Rather, resources should be expended proportionate to the demand for biometrics in the cloud. Meaning, there is a positive correlation between resources available and demand cloud-based biometrics. It has been determined that 85% of idle resources in the cloud are totally unacceptable. Also, the biometric algorithms need elasticity to fit the scalability requirements of the cloud.
6. *Legalities and privacy rights could become grave issues*: Biometrics is one of those technologies that will always be prone to claims of violations of civil liberties, loss of personal freedom, etc. This fear will only be heightened as biometrics move into the cloud because the biometric templates will be literally held at the hands of the hosting provider. In addition, grave legal concerns could take place to establish the clear separation between the biometric templates and other information and data that are stored at the hosting provider, although technically they may be separate as they will be stored on different servers.
7. *Small-scale biometrics applications are best suited*: It has been determined at the present time that small-scale biometric implementations are much better suited as a cloud-based offering than the much larger-scale biometric applications. The former includes such things as the template creation process and verification scenarios. The major culprit inhibiting the growth of large-scale biometrics in the cloud is that the mathematical algorithms are not yet optimized for on-demand, elastic virtualization.

INDEX

Page numbers followed by f and t indicate figures and tables, respectively.

A

Ability to verify rate (AVR), 19
Acceptability
 facial recognition, 84
 fingerprint recognition, 61
 iris recognition, 92
 keystroke recognition, 112
 retinal recognition, 100
 signature recognition, 108
 vein pattern recognition, 73
 voice recognition, 103
Access control list (ACL), 244, 252–253. *See also* Cryptography
Access control rights, 238
Access rights, 238
Accuracy, facial recognition, 83
Acoustic pulses, 34
Active–active failover system, 302
Active–passive failover system, 302
Active sensors with structured lighting, 35
Administration decision making, 150–154. *See also* C-level executive
 biometric template adaptation control, 152
 data transmission, 153–154
 end users, privileges to, 153
 reporting and control, 152–153
 security threshold values, establishment of, 152
 system mode adjustment control, 153
Advanced Encryption Standard (AES) algorithm, 217. *See also* Symmetric cryptography
AFIS (Automated Fingerprint Identification System), 10, 44
Aging, 39. *See also* Sensors
Airports, iris recognition at, 90
Anonymous attacker, 296. *See also* Security threats to cloud computing
ANSI/INCITS 398-2008, 27
ANSI INCITS BioAPI Standards, 29
ANSI INCITS Data Format Standards, 25
ANSI/NIST ITL for law enforcement, 25
Application classifiers. *See also* Biometrics project management guide
 attended *vs.* unattended biometric system, 160
 constrained *vs.* unconstrained biometric system, 160
 cooperative users *versus* uncooperative users, 159
 habituated users *vs.* nonhabituated users, 159–160
 isolated *vs.* integrated biometric system, 160
 overt *vs.* covert biometric system, 159
 private users *vs.* public users, 160
 scalable *vs.* nonscalable biometric system, 160–161
Application layer, 197
Application Programming Interface (API), 318
Application server, 243–244. *See also* Cryptography

Arches, fingerprint, 54
ASP.Net, 318
Asset scaling, 272–273. *See also* Cloud computing
Asymmetric cryptography. *See also* Cryptography
  advantages, 220
  disadvantages of, 220–221
  mathematical algorithms for
    Diffie Hellman asymmetric algorithm, 221–222
    elliptical wave theory algorithm, 222
    public key infrastructure (PKI), 223–224
    RSA algorithm, 221
  primary goal of, 218
  and symmetric cryptography, differences, 219–220
Asymmetric key cryptography, 208–209, 217–218
Asymmetric workload distribution, 300
Asynchronous biometric multimodal systems, 127–128
Attendance, time and, 42–43. *See also* Biometrics
Attended biometric system, 160
Audio feedback, 166
Audit monitor, 301. *See also* Cloud computing
Authentication, 11, 256
  keys, 229
  weak, 298
Authenticity, 294. *See also* Security threats to cloud computing
Authorization, 11
  insufficient, 298. *See also* Cloud computing
Automated Fingerprint Identification System (AFIS), 10, 44, 53, 320
Availability, IT assets/resources, 275–276, 294. *See also* Security threats to cloud computing
Availability rate metric, 307
Availability risks, VPN, 237

B

Backup and recovery of database, 143
Backward compatibility, facial recognition, 83
Bandwidth
  constraints, 278
  optimization, 317
Behavioral biometrics, 1–4, 7–8, 49–50, 311. *See also* Biometrics
  defined, 49
Behavioral model creation, 137
Binary mathematical file, 12, 13
Binary search trees, 141
Binning techniques, 146
BioAPI, 28
  benefits, 28–29
  components, 29
  function of, 28
BioAPI 2.0, 29
Biocryptography. *See also* Cryptography
  cipher biometric template, 256
  ciphertext, 256
  in client–server biometric system, 259–260
  defined, 255
  fingerprint and iris templates, protecting, 257–261
  in hosted biometrics environment, 261
  IPSec, 262–264
  keys, 256–257
  plaintext biometric template, 256
  in single biometric system, 257–259
  and VPN, 261–262
Biometric and token technology application modeling language (BANTAM), 171
Biometric attributes, 173
Biometric data interchange formats, 23–26
Biometric devices, 151
Biometric information records (BIR), 27
Biometric process, 12

Biometric raw image
  postprocessing of, 138
  preprocessing of, 133–135
    alignment, 134
    detection, 134
    normalization process, 135
    segmentation, 134
Biometric recognition server, 144
Biometrics
  behavioral, 1–4, 7–8
  biometric template
    defining, 12
    mathematical files, 12–13
  C-level executive, 17–18
  database, defined, 139
  definition, 5, 6
  DNA recognition, 8–9
  earlobe recognition, 9
  gait recognition, 9
  identification of individual, 5–6
  illustrated, biometric process, 16–21
  key performance indicators (KPI), 18–23
  market segments
    law enforcement, 43–45
    logical access control, 40–41
    physical access control, 41–42
    surveillance, 45–47
    time and attendance, 42–43
  myths behind, 13–14
  physiological, 1–4, 7–8
  recognition, 7
    granular components of, 9–11
  sensors, disadvantages of, 39
  sensors, review of, 31–39
    multispectral fingerprint scanners, 34
    optical scanners, 32–33
    solid-state sensors, 33–34
    temperature differential sensors, 34
    touchless fingerprint sensors, 35–39
    ultrasound sensors, 34
  US federal government biometrics and KPI
    biometric data interchange formats, 23–26
    biometric technical interface standards, 28–29
    Common Biometric Exchange Format Framework (CBEFF), 26–27
  US federal government biometric testing standards, 30–31
  verification and enrollment templates, 14–15
Biometrics as a service (BaaS), 261, 263
Biometric searching, 147
Biometrics in cloud, 310–321. *See also* Cloud computing
  advantages, 325–327
  aspects of, 311
  definition, 310
  disadvantages, 327–328
  networks, 316
  small-business owner, 311–315
  storage, 315–316
  virtual servers, 315
Biometrics project management guide, 154–202. *See also* C-level executive
  application classifiers, 159–161
  biometric interface, 187–188
  biometric networking topologies, 192–193
  biometric system specifications, 172–173
  client-server network topology, 197–198
  COTS-based biometric systems, 161–162
  database management system, 181–183
  data packets, 193–194
  data packet subcomponents, 194–195
  data packet switching, 195
  data storage subsystem design, 180–181

end user/administrator training, 188
feasibility study, 158–159
feedback system, 166–167
human equation, 164–165
information processing architecture, 178–179
infrastructure assessment, 163–164
middleware, 187
networking scenario, 192
network processing loads, 191–192
network protocols, 195–196
network traffic collisions, 201–202
open-ended biometric system, 162–163
operational architecture design, multimodal systems, 177–178
peer-to-peer network topology, 198–199
population dynamics, 167–169
proprietary biometric system, 162
routers, 199–200
routing tables, 200–201
security threshold, determining, 183–184
storage/matching combinations, 175–177
subsystem analysis/design, 179–180
subsystem implementation/testing, 184–185
system architectural/processing designs, 174–175
system concepts/classification schemes, 155–157
system deployment/integration, 185–187
system design/interoperability factors, 161
system maintenance, 188–189
system networking, 191
system reports and logos, 190–191
system requirements
analysis, 169–170
analysis and regulation, 170–171
documentation, 171–172
elicitation, 170
validation, 172
TCP/IP, 196–197
upgradation of existing system, 157–158, 189–190
Biometric subsystem-level testing, 184
Biometrics vendors, 314, 315
Biometric system acceptance testing, 185
Biometric system design constraints, 173
Biometric system external interfaces, 172
Biometric system functionality, 172
Biometric system integration-level testing, 184
Biometric system performance, 173
Biometric system testing, 173
Biometric systemwide-level testing, 184
Biometric technical interface standards, 28–29
Biometric technologies
choice of, 51–52
DNA recognition
advantages/disadvantages, 116
features of, 114–116
earlobe recognition
advantages/disadvantages, 121–122
features of, 120–121
facial recognition, 76–86
advantages/disadvantages, 82–84
applications of, 84–86
effectiveness of, 79–80
features of technology, 77–79
Razko Security adding face recognition technology (case study), 85–86
techniques of, 80–82
fingerprint recognition, 52–65
advantages/disadvantages, 59–62
biometric system for Iraqi Border Control (case study), 63–64
features, 54

market applications of, 62–65
matching algorithm, 58–59
methods of fingerprint collection, 57–58
process of, 55–56
quality control checks, 56
tracking millions of inmates and visitors at U.S. Jails (case study), 64–65
of future, 113–122
gait recognition
advantages/disadvantages, 118–119
process behind, 117–118
hand geometry recognition, 65–69
advantages/disadvantages, 67–69
enrollment process, 66–67
Yeager Airport (case study), 68–69
iris, 87
physiological structure of, 87–88
iris recognition
advantages/disadvantages, 90–93
Afghan Girl—Sharbat Gula (case study), 92–93
market applications of, 90
methods, 89
keystroke recognition, 109–113
advantages/disadvantages, 111–113
process of, 110–111
palm print recognition
advantages/disadvantages, 76
science behind, 76
physical and behavioral, 49–50
physical and behavioral biometrics, differences, 50–51
retina, 93–94
physiology of, 95–96
retinal recognition
advantages/disadvantages, 98–100
process, 96–97
security application, 51
signature recognition, 104–109
advantages/disadvantages, 107–109
features of, 105–107
signature and signature recognition, differences, 105
vein pattern recognition (VPR), 69–75
advantages/disadvantages, 72–75
components of, 71
techniques used, 71–72
Yarco Company (case study), 74–75
voice recognition
advantages/disadvantages, 102–103
factors affecting, 101–102
market applications, 103–104
working of, 100–101
Biometric technology system architecture, 125–126
Biometric template
adaptation control, 152. *See also* Administration decision making
defining, 12
free flow of, 27
mathematical files, 12–13
storage of, 280
Blade servers, 142
Block ciphers, 211–212
Boolean mathematics, 89
Bottom lighting, 72
Broken-up biometric template, 194

C

Caesar key, 211
Caesar methodology, 209–210
CCD (charged couple device), 33, 35–36
CCTV technology, 128
Centralized database, 144
Central server, 194
Certificate authority (CA), 223, 225
Certificate management messages, 253

Certification request syntax standard, 227. *See also* Public Key Cryptography Standards (PKCS)
Chaining certificates, 225
Charged couple device (CCD), 33, 35–36
Chosen-plaintext attack, 210. *See also* Cryptographic attacks
Cipher, 208
Cipher biometric template, 256. *See also* Biocryptography
Cipher block chaining, 212–213
Ciphertext, 207–208, 230, 231, 232, 256. *See also* Biocryptography; Cryptography
Ciphertext-only attack, 210. *See also* Cryptographic attacks
Cleartext, 207, 256
C-level executive, 17–18, 22
  administration decision making, 150–154
    biometric template adaptation control, 152
    data transmission, 153–154
    end users, privileges to, 153
    reporting and control, 152–153
    security threshold values, establishment of, 152
    system mode adjustment control, 153
  biometrics project management guide, 154–202
    application classifiers, 159–161
    biometric interface, 187–188
    biometric networking topologies, 192–193
    biometric system specifications, 172–173
    client–server network topology, 197–198
    COTS-based biometric systems, 161–162
    database management system, 181–183
    data packets, 193–194
    data packet subcomponents, 194–195
    data packet switching, 195
    data storage subsystem design, 180–181
    end user/administrator training, 188
    ergonomic issues, 165–166
    feasibility study, 158–159
    feedback system, 166–167
    human equation, 164–165
    information processing architecture, 178–179
    infrastructure assessment, 163–164
    middleware, 187
    networking scenario, 192
    network processing loads, 191–192
    network protocols, 195–196
    network traffic collisions, 201–202
    open-ended biometric system, 162–163
    operational architecture design, multimodal systems, 177–178
    peer-to-peer network topology, 198–199
    population dynamics, 167–169
    proprietary biometric system, 162
    routers, 199–200
    routing tables, 200–201
    security threshold, determining, 183–184
    storage and matching combinations, 175–177
    subsystem analysis and design, 179–180
    subsystem implementation and testing, 184–185
    system architectural and processing designs, 174–175
    system concepts/classification schemes, 155–157

system deployment and integration, 185–187
system design/interoperability factors, 161
system maintenance, 188–189
system networking, 191
system reports and logos, 190–191
system requirement analysis and regulation, 170–171
system requirements documentation, 171–172
system requirements elicitation, 170
system requirements validation, 172
systems requirements analysis, 169–170
TCP/IP, 196–197
upgradation of existing system, 157–158, 189–190
biometric technology system architecture, 125–126
data storage, 139–145
backup and recovery of database, 143
database configuration, 144–145
database search algorithms, 141–143
search and retrieval techniques, 140–141
goal of, 18
sensing and data acquisition, 126–133
multimodal biometric systems, 127–128, 129–131, 131–133
single sign-on solutions, 128–129
signal and image processing
biometric raw image, preprocessing of, 133–135
data compression, 138–139
feature extraction, 137–138
image enhancement, 136–137
postprocessing, 138
quality control checks, 135–136
template matching, 145–147
threshold decision making, 147–150
Client database, 144
Client-server biometric system, 259–260. *See also* Biocryptography
Client-server network topology, 197–198, 259. *See also* Biometrics project management guide
Closed circuit television (CCTV) camera system, 78, 79
Closed-source database model, 315, 316
Cloud, defined, 269–270
Cloud computing, 261
asset scaling, 272–273
availability/reliability, 275–276
biometrics in cloud, 310–321
advantages, 325–327
aspects, 311
definition, 310
disadvantages, 327–328
networks, 316
small-business owner, 311–315
storage, 315–316
virtual servers, 315
challenges/risks of, 276–277
characteristics of, 280–292
elasticity, 283
measured usage, 283
on-demand usage, 281
resiliency, 284
resource pooling, 282
ubiquitous access, 282
cloud, defined, 269–270
cloud-computing delivery models
infrastructure as a service (IaaS), 284–286
platform as a service (PaaS), 286–288
software as a service (SaaS), 288–289
cloud-computing deployment models
community cloud, 291
hybrid cloud, 292
private cloud, 291
public cloud, 289–290
concepts, 268

cost metrics, 303–307
cloud storage device usage, 305–306
cost management, 306–307
network usage, 304
server usage, 305
distinctions, 270
example, in details, 321–325
IT resource, 270–271
mechanisms, 299–303
audit monitor, 301
failover system, 301–302
hypervisor, 302
load balancers, 300
pay-per-use monitor, 301
resource clustering, 302–303
on-premise component, 271
overview, 267–268
proportional costs, 274
scalability, 271–272, 274–275
security risk/challenges, 277–280
compliance and legal issues, 279–280
limited portability, 278–279
reduced operational governance, 278
security threats to, 292–299
anonymous attacker, 296
authenticity, 294
availability, 294
confidentiality, 294
denial of service, 298
insufficient authorization, 298
integrity, 294
malicious insider, 297
malicious intermediary, 298
malicious service agent, 296–297
overlapping trust boundaries, 299
security controls, 295
security mechanisms, 296
security policies, 296
security risk, 295
threat, 295
traffic eavesdropping, 297
trusted attacker, 297
virtualization attack, 299
vulnerability, 295
service quality mechanisms
service availability metrics, 307
service performance metrics, 308–309
service reliability metrics, 307–308
service resiliency metrics, 309–310
service scalability metrics, 309
SLA agreements, 276
terminology, 268
Cloud providers, 272
Cloud service contracting, 306
Cloud service design/development, 306
Cloud service offerings, 306
Cloud service operations, 306
Cloud service provisioning, 306
Cloud services deployment, 306
Cloud storage device usage, 305–306
Cluster types, cloud-based. *See* Resource clustering
Cohort matching, 147
Collectability
facial recognition, 83
fingerprint recognition, 60
iris recognition, 91
keystroke recognition, 112
retinal recognition, 99
signature recognition, 108
vein pattern recognition, 72
voice recognition, 103
Combined DNA System (CODIS), 114
Commercialization, 290
Common Biometric Exchange Format Framework (CBEFF), 26–27. *See also* Biometrics
Common file format, 27
Community cloud, 291
Completion time metric, 309
Composite image, 16
Computer hardware, 178

Computer network access, 40
Confidentiality, 294. *See also* Security threats to cloud computing
Confidentiality risk, VPN, 237
Constrained biometric system, 160
Content-aware distribution, 300
Correlation-based matching, 59
Cost, cloud computing
  proportional, 274
Cost-benefit analysis (CBA), 235–236
Cost metrics, 303–307. *See also* Cloud computing
  cloud storage device usage, 305–306
  cost management, 306–307
  network usage, 304
  server usage, 305
Cost savings, 199
COTS (commercial off-the-shelf biometric system)-based biometric systems, 161–162. *See also* Biometrics project management guide
Coupling, defined, 179
Covert surveillance, 46
Credit card information, 324
Critical infrastructures, iris recognition at, 90
Cryptographic attacks
  chosen-plaintext attack, 210
  ciphertext-only attack, 210
  known-plaintext attack, 210
Cryptographic message syntax standard, 227. *See also* Public Key Cryptography Standards (PKCS)
Cryptographic token information format standard, 228. *See also* Public Key Cryptography Standards (PKCS)
Cryptographic token interface standard, 227
Cryptography. *See also* Biocryptography
  access control list (ACL), 252–253
  application server, impacts to, 243–244
  asymmetric cryptography
    disadvantages of, 220–221
    and symmetric cryptography, differences, 219–220
  asymmetric key cryptography, 217–218
  attacks against, 247
  and biocryptography, 205
  biometrics-based VPN, vulnerabilities of, 255
  block ciphers, 211–212
  Caesar methodology, 209–210
  cipher block chaining, 212–213
  ciphertexts, 207–208
  cryptographic attacks
    chosen-plaintext attack, 210
    ciphertext-only attack, 210
    known-plaintext attack, 210
  database server, impacts to, 244–245
  digital certificates, 224
  encryption/decryption, 207
  end users and employees, 239–240
  firewall, impacts to, 245
  hashing function, 217
  initialization vectors (IV), 212
  internet drafts, 253–255
  IP tunneling, 232–234
  key distribution center (KDC), 215–216
  keys and public private keys, 218–219
  Lightweight Directory Access Protocol (LDAP), 226–227
  mathematical algorithms for asymmetric cryptography
    Diffie Hellman asymmetric algorithm, 221–222
    elliptical wave theory algorithm, 222
    public key infrastructure (PKI), 223–224
    RSA algorithm, 221
  mathematical algorithms with symmetric cryptography

Advanced Encryption Standard (AES) algorithm, 217
Digital Encryption Standard (DES) algorithm, 216
International Data Encryption Algorithm (IDEA), 217
Needham-Schroder algorithm, 216
Triple Digit Encryption Standard algorithm (3DES), 216
message digests and hashes, 230–231
message scrambling/descrambling, 206–207
mobile VPN, 234
network requirements, 240–241
overview, 205–206
PKI
policies/rules, 225–226
working of, 224–225
polyalphabetic encryption, 210–211
Public Key Cryptography Standards (PKCS), 227–228
public keys and private keys
parameters, 228–229
securing, 230
public/private key exchanges management, 251–252
ISAKAMP/Oakley Protocol, 251
Manual Key Exchanges, 251
Simple Key Interchange Protocol (SKIP), 251
security policies, 230
security vulnerabilities of hashes, 231
servers, 229
symmetric key cryptography, disadvantages of, 213–215
symmetric key systems/ asymmetric key systems, 208–209
VPN, 232, 234–235
building, 241–242
cost-benefit analysis, 235–236
implementing, 236–238, 249–251
security policy, components of, 238–239
testing, 246–248
web server, impacts to, 242–243
Cyber attack, 295. *See also* Cloud computing
Cyclical redundancy checking (CRC) algorithm

D

Data acquisition. *See* Sensing and data acquisition
Database, recovery of, 143
Database clustering, 303. *See also* Resource clustering
Database configuration, 144–145
Database designers, 181
Database infrastructure, 164
Database management system (DBMS), 181–183
Database search algorithms, 141–143. *See also* C-level executive
binary search trees, 141
blade servers, 142
hashing, 141
parallel searching, 141
sequential searching, 141
virtual servers, 142
Database server, 244–245. *See also* Cryptography
Data compression, 138–139
Data elements, 27
Data encryption keys, 229
Data formats, 27
Data Format Standards, 25
Data packet, 193–194
subcomponents
4 bits, 194
8 bits, 194
16 bits, 195
32 bits, 195
switching, 195
Data packet optimization, 200

Data security, 277
Data storage, 139–145
  backup and recovery of database, 143
  database configuration, 144–145
  database search algorithms, 141–143
  search and retrieval techniques, 140–141
  subsystem design, 180–181
Data transmission, 153–154. *See also* Administration decision making
Data warehousing techniques, 244
Decentralized database, 144
Decision, defined, 148
Decision-level fusion, 177
Decryption, 207, 256
  defined, 207
Degradation, facial recognition biometric system, 83
Delivery models, cloud-computing. *See also* Cloud computing
  infrastructure as a service (IaaS), 284–286
  platform as a service (PaaS), 286–288
  software as a service (SaaS), 288–289
Denial of service, 298
Deployment models, cloud-computing. *See also* Cloud computing
  community cloud, 291
  hybrid cloud, 292
  private cloud, 291
  public cloud, 289–290
Design phase, cloud-computing infrastructure, 310
DH algorithm. *See* Diffie Hellman asymmetric algorithm
Diffie Hellman asymmetric algorithm, 221–222. *See also* Asymmetric cryptography
Diffie-Hellman Proof of Possession Mathematical Algorithms, 254
Diffused illumination, 71
Digital certificates, 223, 224, 252. *See also* Cryptography
  issuer name, 224
  public key, 224
  serial number, 224
  signature algorithm identifier, 224
  subject name URL, 224
  version number, 224
Digital criminal records, 45
Digital Encryption Standard (DES) algorithm, 216. *See also* Symmetric cryptography
Digital signatures, 105, 223
Dimensional touchless finger imaging, 35
Direct illumination, 71–72
Distributed processing, 179
DNA code, 1, 2, 8, 87
DNA loci, 115–116
DNA profiling, 114
DNA recognition, 8–9. *See also* Biometrics; Biometric technologies
  advantages/disadvantages
    acceptability, 116
    collectability, 116
    performance, 116
    resistance to circumvention, 116
    uniqueness, 116
  features of, 114–116
Doppler radar, 117–118
Dummy prints, 62
Dynamic signature recognition, 105
Dynamic threshold, 149
Dynamic time warping, 118

E

Earlobe recognition, 9. *See also* Biometrics; Biometric technologies
  advantages/disadvantages
    acceptability, 122
    collectability, 121

performance, 121
permanence, 121
resistance to circumvention, 122
uniqueness, 121
universality, 121
features of, 120–121
E-commerce, 288
Eigenears, 120
Eigenfaces, 81, 120
Elastic bunch graph matching (EBGM), 82
Elasticity, cloud-computing infrastructure, 283
Electric currents, 34
Elliptical Digital Certificate Algorithm, 253
Elliptical wave theory, 222
Elliptical wave theory algorithm, 222. *See also* Asymmetric cryptography
Encapsulation, 232
Encryption, 207
algorithm, 208
defined, 207
Endogenous data, 146
End user(s)
and employees, 239–240. *See also* Cryptography
privileges to, 153. *See also* Administration decision making
training, 39
End user/administrator training, 188
Energy level normalization process, 135
Enrollment, 16
templates, 14, 137
Enterprise solutions middleware, 187
Entropy, 228
Environment design, 174
E-passport, 319
Equal error rate (ERR), 19, 20, 21
Ergonomics, 52
Execution efficiency, 179
Exogenous data, 146
Extended certificate syntax standard, 227. *See also* Public Key Cryptography Standards (PKCS)
Extensibility, 27
EyeDentification System 7.5, 96

F

Facial recognition, 13, 35, 45, 76–86, 128, 190. *See also* Biometric technologies
advantages/disadvantages, 82–84
acceptability, 84
collectability, 83
performance, 83
permanence, 83
resistance to circumvention, 84
uniqueness, 83
universality, 82
applications of, 84–86
effectiveness of, 79–80
features of technology, 77–79
Razko Security adding face recognition technology (case study), 85–86
techniques of, 80–82
Failover system, 301–302. *See also* Cloud computing
Failure recovery, 247
Failure to enroll rate (FER), 19
False acceptance rate (FAR), 19, 20, 21, 183
False accept rate, 130
False match rate (FMR), 132
False nonmatch rate (FMNR), 132
False rejection rate (FRR), 19, 20, 21
Fault tolerance analysis, 246
Feature extraction, 137–138
Feature-level fusion
feature normalization, 177
unimodal feature fusion, 177
Feature normalization, 177

Feedback loop mechanism, 135
Feedback system, 166–167. *See also* Biometrics project management guide
Fiberlink, 243
Fibrovascular tissue, 88
File transfer protocol (FTP), 277
  networks, 250
Filtering, 146
Financial transactions, fingerprint recognition in, 62
Fine tuning, 189
Fingerprint, 2
  and iris templates, protecting, 257–261. *See also* Biocryptography
  latent, 327
  minutiae of, 3
  for verification/identification, 53
Fingerprint recognition, 13, 32, 37, 52–65, 258. *See also* Biometric technologies
  advantages/disadvantages, 59–62
    acceptability, 61
    collectability, 60
    performance, 61
    permanence, 60
    resistance to circumvention, 61
    uniqueness, 60
    universality, 60
  biometric system for Iraqi Border Control (case study), 63–64
  features
    arches, 54
    lakes, 54
    loops, 54
    prints/islands, 54
    spurs, 54
    whorls, 54
  market applications of, 62–65
    in financial transactions, 62
    forensics, 62
    government benefits, administration of, 62
    physical and logical access, 62–63
  matching algorithm, 58–59
    correlation-based matching, 59
    minutiae-based matching, 59
    ridge feature matching, 59
  methods of fingerprint collection
    live scan sensors, 57–58
    offline scanning methods, 57–58
    sweep method, 58
    touch method, 58
  process of, 55–56
  quality control checks
    area, 56
    dynamic range, 56
    fingerprint pixels, 56
    frames per second, 56
    geometric accuracy, 56
    image quality, 56
    resolution, 56
  tracking millions of inmates and visitors at U.S. Jails (case study), 64–65
Firewall, 245. *See also* Cryptography
Forensic automatic speaker recognition (FASR), 100
Forensics, 62
Forensics as a service, 327
Fresh virtual instances, 286
Functional architecture
  environment design, 174
  hardware design, 174
Fundus cameras, 95
Fusion data, 25

G

Gabor wavelet mathematics, 82
Gabor wavelet theory mathematics, 13
Gait recognition, 9. *See also* Biometrics; Biometric technologies
  advantages/disadvantages, 118–119
    acceptability, 119
    collectability, 119
    performance, 119
    permanence, 119

resistance to circumvention, 119
uniqueness, 119
universality, 119
process behind
direct matching, 118
Doppler radar, 117–118
dynamic shapes, 117
dynamic time warping, 118
HMM, 118
holistic analysis, 118
machine vision, 117
model-based analysis, 118
static shapes, 117
thermography, 118
Garbage in–garbage out, 50
Gaussian mixture models (GMM), 102
Global statistical profiles, 111
Graphical user interface (GUI), 29, 166

H

Habituated users, 159–160
Hackers, 276, 278
Hamming distances, 89
Hand geometry recognition, 2, 13, 37, 65–69. *See also* Biometric technologies
advantages/disadvantages, 67–69
acceptability, 68
collectability, 67
performance, 67
permanence, 67
resistance to circumvention, 68
uniqueness, 67
universality, 67
enrollment process, 66–67
Yeager Airport (case study), 68–69
Hardware design, 174
Hashes
defined, 217
security vulnerabilities of, 231
Hashing, 141
Hashing function, 217
Hemoglobin, 70
Hidden Markov models (HMM), 51, 102, 118
HMM (Hidden Markov models), 51, 102, 118
Horizontal IT asset scaling, 272, 273
Hosted biometrics environment, 261
Hosting providers, 261, 310, 313, 315, 325
Human equation, 164–165. *See also* Biometrics project management guide
Human to machine interface, 180
Hybrid biometric system, 130
Hybrid cloud, 292
Hybrid method, 149
Hygiene, fingerprint biometric systems, 61
Hypertext transport protocol (HTTP), 242
Hypervisor, 302. *See also* Cloud computing

I

Identification applications, 10
Identification matching, 146, 147
Identification of individual, 5–6
Illumination, diffused/direct, 71–72
Image enhancement, 136–137
Inbound network usage metric, 304
Indexing, biometric information, 146
Individual threshold, 149
Information processing architecture
computer hardware, 178
distributed processing, 179
parallel processing, 178–179
Infrastructure as a service (IaaS), 284–286, 314. *See also* Cloud computing
Infrastructure assessment, 163–164
Initialization vectors (IV), 212
Inmates and visitors at U.S. Jails, tracking millions of (case study), 64–65
Input ports, 201

Instance starting time metric, 308
Insufficient authorization, 298
Integrated Automated Fingerprint Identification System (IAFIS), 44, 45, 53
Integrated biometric system, 160
Integrity, 256
  cloud computing, 294. *See also* Cloud computing
Integrity risk, VPN, 237
Interface, biometric, 187–188
International Civil Aviation Organization (ICAO), 25, 84
International Data Encryption Algorithm (IDEA), 217. *See also* Symmetric cryptography
Internet drafts, 253–255
Internet header length, 194
Internet protocol, 196
Internet service providers (ISP), 248, 267, 271
  biometric business processes, changes in, 327
  business processes, 327
  lag time for, 327
  legalities and privacy rights, 328
Internet X.509 Public Key Infrastructure Extending Trust, 254
Internet X.509 Public Key Infrastructure Operational Protocols, 254
Internet X.509 Public Key Infrastructure PKIX Roadmap, 254
Internet X.509 Public Key Infrastructure-Qualified Certificates, 254
Internet X.509 Public Key Infrastructure Time Stamp Protocol, 254
Interoperability factors, 161
Interoperability of sensor, 39
Interstate Index, 45
Intracloud WAN usage metric, 304
I/O data-transferred metric, 306
IPSec, 262–264. *See also* Biocryptography
IP tunneling, 232–234, 262, 263
Iraqi Border Control, biometric system for (case study), 63–64
Iridian Technologies, 87
Iris, 87
  physiological structure of, 87–88
  primary purpose of, 88
IrisCode, 89
Iris recognition, 13, 37, 192. *See also* Biometric technologies
  advantages/disadvantages, 90–93
    acceptability, 92
    collectability, 91
    performance, 91
    permanence, 91
    resistance to circumvention, 92
    uniqueness, 91
    universality, 91
  Afghan Girl-Sharbat Gula (case study), 92–93
  market applications of
    airports, 90
    critical infrastructures, 90
    logical access entry, 90
    military checkpoints, 90
    physical access entry, 90
    seaports, 90
    time and attendance, 90
Iris templates, protecting, 257–261. *See also* Biocryptography
ISAKAMP, 251. *See also* Public/private key exchanges management
ISO/IEC 19784, 29
ISO/IEC 19785, 27
ISO/IEC 19795-1:2006, 30
ISO/IEC 19795-2:2007, 30
ISO/IEC 19795-4:2008, 30
ISO/IEC CD 19795-5, 30
ISO/IEC CD 19795-6, 30
ISO/IEC CD 19795-7, 30
ISO/IEC TR 19795-3:2007 Technical Report, 30
Isolated biometric system, 160

IT assets, virtualized, 285
IT environment, controlled, 271
IT resource, 270–271. *See also* Cloud computing

K

Kerberos, 215
Key distribution center (KDC), 215–216
Key encryption keys, 229
Key escrow, 230
Key performance indicators (KPI), 17, 18–21, 22–23. *See also* Biometrics
 US federal government biometrics and
  biometric data interchange formats, 23–26
  biometric technical interface standards, 28–29
  Common Biometric Exchange Format Framework (CBEFF), 26–27
Key recovery, 230
Keys, 256
 biocryptography, 256–257
 and public private keys, 218–219
Keyspace, 257
Keystroke recognition, 13, 38, 39, 109–113. *See also* Biometric technologies
 advantages/disadvantages, 111–113
  acceptability, 112
  collectability, 112
  performance, 112
  permanence, 112
  resistance to circumvention, 112
  uniqueness, 112
  universality, 112
 process of, 110–111
Known-plaintext attack, 210. *See also* Cryptographic attacks
KPI. *See* Key performance indicators (KPI)

L

Lakes, fingerprint, 54
Large data-set cluster, 303. *See also* Resource clustering
Laser, 36
Latent fingerprints, 327
Law enforcement, 25, 43–45, 320. *See also* Biometrics
 advantages, 44
LDAP. *See* Lightweight Directory Access Protocol (LDAP)
Legalities and privacy rights, 328
Letter sequencing, 209
Light-emitting diodes (LED), 32, 66
Lightweight Directory Access Protocol (LDAP), 223, 226–227
Linear discriminant analysis (LDA), 81, 82
Link layer, 196
Live sample data, 126
Live scan sensors, 57–58
Load balancers, 300. *See also* Cloud computing
Local area network (LAN), 259
Local statistical profiles, 111
Logical access control, 40–41
Logical access entry, iris recognition at, 90
Long-term form, 5
Loops, fingerprint, 54
Lossless data compression, 139

M

Machine vision, 117
Malicious insider, 297
Malicious intermediary, 298. *See also* Cloud computing
Malicious service agent, 296–297
Manual key exchanges, 251. *See also* Public/private key exchanges management
Manual labor jobs, 165

Market segments, biometric. *See also* Biometrics
  law enforcement, 43–45
  logical access control, 40–41
  physical access control, 41–42
  surveillance, 45–47
  time and attendance, 42–43
Massively parallel processing (MPP), 242
Matching algorithm, fingerprint recognition, 58–59
Mathematical algorithms. *See also* Cryptography
  with symmetric cryptography
    Advanced Encryption Standard (AES) algorithm, 217
    Digital Encryption Standard (DES) algorithm, 216
    International Data Encryption Algorithm (IDEA), 217
    Needham-Schroder algorithm, 216
    Triple Digit Encryption Standard algorithm (3DES), 216
Mathematical algorithms for asymmetric cryptography
  Diffie Hellman asymmetric algorithm, 221–222
  elliptical wave theory algorithm, 222
  public key infrastructure (PKI), 223–224
  RSA algorithm, 221
Measured usage, cloud-computing infrastructure, 283
Message descrambling, 206–207
Message digests, 231
  and hashes, 230–231
Message integrity, 217
Message scrambling, 206–207
Middleware, 187. *See also* Biometrics project management guide
Military checkpoints, iris recognition at, 90
Minutiae-based matching, 59
Minutiae of fingerprint, 3
Mitochondrial DNA, 115
Mobile VPN, 234
Model-based facial recognition techniques, 81
Model requirements, 170, 171
Monoalphabetic cipher, 209
Morse Code, 110
Multimodal biometric systems, 127–128. *See also* Sensing and data acquisition
  challenges with, 131–133
  implementing, 129–131
Multiregional compliance and legal issues, 279–280
Multispectral fingerprint scanners, 34. *See also* Sensors
Multitenancy model, 282

N

National ID card, 319
National Institutes of Standards and Technology (NIS), 268
Near-infrared (NIR) light, 70
Near-infrared spectroscopy (NIRS), 37, 38
Needham-Schroder algorithm, 216. *See also* Symmetric cryptography
Network and directory infrastructure, 164
Network capacity metric, 308
Networking, 323
Networking topologies, biometric, 192–193
Network processing loads, 191–192. *See also* Biometrics project management guide
Network protocol data packets, 194
Network protocols, 195–196. *See also* Biometrics project management guide
Network traffic collisions, 201–202. *See also* Biometrics project management guide

Network usage, 304. *See also* Cost metrics
Nonhabituated users, 159–160
Nonrepudiation, 256
Nonscalable biometric system, 160–161
Nuclear DNA, 115

O

Oakley Protocol, 251. *See also* Public/private key exchanges management
Obsolescence, 39. *See also* Sensors
Offline scanning methods, 57–58
On-demand storage space allocation metric, 305
On-demand usage, cloud-computing infrastructure, 281
On-demand virtual machine instance allocation metric, 305
On-premise component, 271
Open-ended biometric system, 162–163
Open-source database model, 315, 318
Operating mode, biometric system, 153
Operational architecture, 175
Operational architecture design, multimodal systems, 177–178
Operational phase, cloud-computing infrastructure, 310
Optical network, 243
Optical scanners, 32–33. *See also* Sensors
Optical sensors, 58
Outage duration metric, 307
Outbound network usage metric, 304
Output ports, 201
Overlapping trust boundaries, 278
Overt surveillance, 46

P

Packet switching, 199
Padding, 212
Palm print recognition. *See also* Biometric technologies
  advantages/disadvantages, 76
    acceptability, 76
    collectability, 76
    performance, 76
    resistance to circumvention, 76
  science behind, 76
Palm print technology, 76
Paperless documents, 105
Parallel processing, defined, 178
Parallel searching, 141
Passive/assisted stereo sensing, 35
Passive feedback, 166
Password, 40
Password-based cryptography standard, 227. *See also* Public Key Cryptography Standards (PKCS)
Pay-per-use monitor, 301. *See also* Cloud computing
PC clustering, 243
Peers, defined, 199
Peer-to-peer biometric network, 197f
Peer-to-peer network topology, 198–199. *See also* Biometrics project management guide
Performance
  facial recognition, 83
  fingerprint recognition, 61
  iris recognition, 91
  keystroke recognition, 112
  retinal recognition, 99
  signature recognition, 108
  vein pattern recognition, 72–73
  voice recognition, 103
Permanence
  facial recognition, 83
  fingerprint recognition, 60
  iris recognition, 91
  keystroke recognition, 112
  retinal recognition, 99
  signature recognition, 108
  vein pattern recognition, 72
  voice recognition, 103
Personal identification numbers (PIN), 4

Personal information exchange standard, 228. *See also* Public Key Cryptography Standards (PKCS)
Phasors, 89
Physical access control, 41–42, 251. *See also* Biometrics
Physical access entry, iris recognition at, 90
Physical and behavioral biometrics, differences, 50–51
Physical biometrics, 49–50, 311
Physical infrastructure, 163
Physiological biometrics, 1–4, 7–8
Piezoelectric effect sensors, 34
PKI. *See* Public key infrastructure (PKI)
Plaintext, 207, 256
Plaintext biometric template, 256. *See also* Biocryptography
Platform as a service (PaaS), 286–288, 314. *See also* Cloud computing
Platform infrastructure, 163
Plug and play (BioAPI standard), 28
Polyalphabetic encryption, 210–211
Population dynamics, 167–169. *See also* Biometrics project management guide
  interclass variability, 167
  intraclass variability, 167
Portable storage medium, 144, 145
Post-it syndrome, 4, 40
Postprocessing, biometric raw image, 138
Presentation, defined, 126
Primary cloud, 321
Principal component analysis (PCA), 66, 81, 121
Prints/islands, fingerprint, 54
Privacy, 276
Privacy rights issues, 61
Private cloud, 291
Private key information syntax standard, 227. *See also* Public Key Cryptography Standards (PKCS)
Private keys, 214, 215
Private users, 160
Probe data, 126
Proprietary biometric system, 162
Pseudo-random number generators, 228
Public cloud, 289–290
Public key, 218
Public Key Cryptography Standards (PKCS). *See also* Cryptography
  certification request syntax standard, 227
  cryptographic message syntax standard, 227
  cryptographic token information format standard, 228
  cryptographic token interface standard, 227
  extended certificate syntax standard, 227
  password-based cryptography standard, 227
  personal information exchange standard, 228
  private key information syntax standard, 227
  selected attribute types, 227
Public key infrastructure (PKI), 223–224. *See also* Asymmetric cryptography
  policies/rules, 225–226
  working of, 224–225
Public/private key exchanges management, 251–252
  ISAKAMP/Oakley Protocol, 251
  manual key exchanges, 251
  Simple Key Interchange Protocol (SKIP), 251
Public private keys, 218–219
  parameters, 228–229
  securing, 230
Public users, 160
Pupillary zone, 88
Pyroelectric material, 34

Q

Quality assurance lifecycle, 185
Quality control checks, 135–136
  in feedback loop, 135
  as input mechanism, 136
  as predictive mechanism, 135–136
Quality control score, 136
Quality metrics, components
  attainable, 307
  availability, 306
  comparability, 307
  performance, 306
  quantifiable, 306
  reliability, 306
  repeatability, 307
  resiliency, 306
  scalability, 306
Quality of service, 194
Quality score, 149

R

Random early detection, 202
Random number generators, 228
Rank-based technique, 148
Rank requirements, 170, 171
Rate of return on investment (ROI), 18
Raw data
  live sample data, 126
  probe data, 126
Raw images of fingerprint, 55
Razko Security adding face recognition technology (case study), 85–86
Recognition, 7. *See also* Biometrics
  granular components of, 9–11
Recovery phase, cloud-computing infrastructure, 310
Re-enrollment, 152
Reflection-based touchless finger imaging, 35
Registration authority (RA), 223, 225
Reliability, cloud computing, 275–276
Reliability rate metric, 308
Reof key, 229
Reporting and control mechanism, 152–153
Requests for information (RFI), 157
Requests for proposals (RFP), 157
Requirement, defined, 169
Reserved virtual machine instance allocation metric, 305
Resiliency, cloud-computing infrastructure, 284
Resistance to circumvention
  facial recognition, 84
  fingerprint recognition, 62
  iris recognition, 92
  keystroke recognition, 112
  retinal recognition, 100
  signature recognition, 108
  vein pattern recognition, 73
  voice recognition, 103
Resource clustering, 302–303. *See also* Cloud computing
  database clustering, 303
  large data-set cluster, 303
  server cluster, 303
Resource pooling, cloud-computing infrastructure, 282
Response time metric, 309
Retina, 93–94
  physiology of, 95–96
Retinal recognition. *See also* Biometric technologies
  advantages/disadvantages, 98–100
    acceptability, 100
    collectability, 99
    performance, 99
    permanence, 99
    resistance to circumvention, 100
    uniqueness, 99
    universality, 99
  process
    image/signal acquisition and processing, 96
    matching, 96
    representation, 96
    scans, poor quality of, 97

Retinal scanning devices, 95
Retinal scanning technology, 98
Retinal structure, 2
Return on investment (ROI), 159
Ridge feature matching, 59
Routers, 199–200
components
input ports, 201
output ports, 201
routing processor, 201
switching fabric, 201
Routing processor, 201
Routing tables, 200–201
RSA algorithm, 221. *See also* Asymmetric cryptography

S

Scalability, cloud computing, 271–272, 274–275
Scalable biometric system, 160–161
Scale/shape normalization process, 135
Scaling down, 273
Scaling in, 272
Scaling out, 272
Scaling up, 273
Scanner, 128
Scope–creep phenomenon, 236
Score-level fusion, 177
Seaports, iris recognition at, 90
Search and retrieval techniques, 140–141
biometric template accuracy requirements, 141
identification, 140
multiple types of biometric data, 141
verification, 140
winnowing errors, 140
Secret key, 219, 231
Secure Shell (SSH) connection, 228
Security policy, 230
VPN, 238–239
Security risk/challenges, cloud computing, 277–280
compliance and legal issues, 279–280
limited portability, 278–279
reduced operational governance, 278
Security threats to cloud computing, 292–299. *See also* Cloud computing
anonymous attacker, 296
authenticity, 294
availability, 294
confidentiality, 294
denial of service, 298
insufficient authorization, 298
integrity, 294
malicious insider, 297
malicious intermediary, 298
malicious service agent, 296–297
overlapping trust boundaries, 299
security controls, 295
security mechanisms, 296
threat, 295
traffic eavesdropping, 297
trusted attacker, 297
virtualization attack, 299
vulnerability, 295
Security threshold values, establishment of, 152. *See also* Administration decision making
Security tool, 40
Self-scalability, 199
Sensing and data acquisition, 126–133. *See also* C-level executive
acceptability, 132
background, 133
circumvention, 132
collectability, 131
contrast, 133
illumination, 133
multimodal biometric systems, 127–128
challenges with, 131–133
implementing, 129–131
performance, 131
single sign-on solutions, 128–129

Sensitivity of matching algorithms, 17
Sensor-level fusion, 177
Sensors, 31. *See also* Biometrics
defined, 31
disadvantages of
aging, 39
end user training, 39
interoperability of sensor, 39
obsolescence, 39
fatigue, 180
multispectral fingerprint scanners, 34
optical scanners, 32–33
review of, 31–39
solid-state, 33–34
temperature differential, 34
touchless fingerprint, 35–39
ultrasound, 34
Sequential searching, 141
Server capacity metric, 308
Server cluster, 303. *See also* Resource clustering
Servers, 229
blade, 142
virtual, 142
Server scalability (horizontal/vertical), 309
Server-side interface, 187
Server usage, 305. *See also* Cost metrics
Service availability metrics, 307
Service level agreement (SLA) agreements, 276. *See also* Cloud computing
Service performance metrics, 308–309
Service reliability metrics, 307–308
Service resiliency metrics, 309–310
Service scalability metrics, 309
Session keys, 228, 229
Short tandem repeat (STR), 114
Short-term form, 5
Side lighting, 72
Signal and image processing. *See also* C-level executive
biometric raw image, preprocessing of, 133–135
data compression, 138–139
feature extraction, 137–138
image enhancement, 136–137
postprocessing, 138
quality control checks, 136–136
Signature recognition, 13, 38, 104–109. *See also* Biometric technologies
advantages/disadvantages, 107–109
acceptability, 108
collectability, 108
performance, 108
permanence, 108
resistance to circumvention, 108
uniqueness, 108
universality, 108
features of, 105–107
signature and signature recognition, differences, 105
Signing keys, 229
Simple Certificate Validation Protocols (SCVP), 255
Simple Key Interchange Protocol (SKIP), 251. *See also* Public/private key exchanges management
Single biometric system, bio, 257–259
Single sign-on solutions (SSO), 41, 128–129
SLA agreements. *See* Service level agreement (SLA) agreements
Small-business owner, 311–315. *See also* Biometrics in cloud
steps by, 324
Small-scale biometrics applications, 328
Small- to medium-sized enterprises (SME), 113
Social Security numbers, 4
Software, 5, 41, 323
Software as a service (SaaS), 261, 288–289, 314, 317. *See also* Cloud computing
Software development, customized, 317, 319

Software development kit (SDK), 130
Solid-state sensors, 33–34, 58. *See also* Sensors
  advantages, 33–34
  types of, 33
Sphincter muscles, 88
Spurs, fingerprint, 54
SQL Server, 272, 273
Stakeholders, 164
Stand-alone iris, 258
Standardization, 24, 279
  facial recognition biometric system, 83
Standard networking protocol, 323
Static threshold, 149
Stereo sensing, passive/assisted, 35
Storage, 315–316, 323. *See also* Biometrics in cloud
  scalability, 309
Storage and matching combinations, 175–177. *See also* Biometrics project management guide
  store on client, match on client architecture, 175
  store on device, match on device architecture, 175
  store on server, match on server architecture, 175
  store on token, match on device architecture, 176
  store on token, match on server architecture, 176
  store on token, match on token architecture, 176
Storage device component metric, 308
Stroma, 88
Subsystem analysis and design, 179–180
Subsystem implementation and testing, 184–185
Surveillance, 45–47. *See also* Biometrics
Sweep method, fingerprint collection, 58
Switching fabric, 201
Symmetric cryptography, 214, 215
  mathematical algorithms with
    Advanced Encryption Standard (AES) algorithm, 217
    Digital Encryption Standard (DES) algorithm, 216
    International Data Encryption Algorithm (IDEA), 217
    Needham-Schroder algorithm, 216
    Triple Digit Encryption Standard algorithm (3DES), 216
Symmetric key cryptography, 208–209
  disadvantages of, 213–215
Symmetric multiprocessing (SMP), 242
Synchronous biometric multimodal systems, 127
System architectural and processing designs, 174–175
  detailed high-level objectives, 174
  high-level goals, 174
  mid-level objectives, 174
System concepts/classification schemes, 155–157
System deployment/integration, 185–187
System design/interoperability factors, 161
System maintenance, 188–189
System mode adjustment control, 153. *See also* Administration decision making
System networking, 191. *See also* Biometrics project management guide
System reports and logos, 190–191
System requirements. *See also* Biometrics project management guide
  analysis, 169–170
  analysis and regulation, 170–171
  documentation, 171–172
  elicitation, 170
  validation, 172
System specifications, biometric, 172–173

T

Tail drop, 202
TCP/IP address, 233
TCP/IP network protocol, 226
Technical interface, 186
Temperature differential sensors, 34. *See also* Sensors
Template adaptation, 152
Template aging, defined, 152
Template matching, 145–147. *See also* C-level executive
Template/model adaptation, 190
Template size, facial recognition biometric system, 83
Terror watch lists, 46
Testing mode, biometric system, 153
Text-dependent voice recognition system, 38
Thermography, 118
Threat, statistical probability of, 295
Threat agents, 292, 296
3-D analysis, 121
3-D sensors, 35, 36
Threshold-based technique, 148
Threshold decision making, 147–150. *See also* C-level executive
Tickets, 215
Time and attendance, iris recognition at, 90
Time and attendance systems, biometrics-based, 42–43. *See also* Biometrics
Top lighting, 71
Touchless fingerprint sensors, 35–39. *See also* Sensors
Touch method, fingerprint collection, 58
Trabecular meshwork, 88
Tracking of individuals
  for suspicious behavior, 46
  for suspicious types of activities, 46
  on watch lists, 46
Traffic eavesdropping, 297. *See also* Cloud computing
Training mode, biometric system, 153
Transfer syntax, 24
Transmission-based touchless finger imaging, 35
Transport layer, 196
Triangulation, 36
Triple Digit Encryption Standard algorithm (3DES), 216. *See also* Symmetric cryptography
Trusted attacker, 297. *See also* Cloud computing
Trusted travel program, 68

U

Ubiquitous access, cloud-computing infrastructure, 282
Ultrasound sensors, 34, 58. *See also* Sensors
Unattended biometric system, 160
Unconstrained biometric system, 160
Unified modeling language (UML), 170, 171
Unimodal feature fusion, 177
Unique feature location, 137
Unique Internet Protocol (IP) address, 316
Uniqueness
  facial recognition, 83
  fingerprint recognition, 60
  iris recognition, 91
  keystroke recognition, 112
  retinal recognition, 99
  signature recognition, 108
  vein pattern recognition, 72
  voice recognition, 102
Universality
  facial recognition, 82
  fingerprint recognition, 60
  iris recognition, 91
  keystroke recognition, 112
  retinal recognition, 99
  signature recognition, 108
  vein pattern recognition, 72
  voice recognition, 103

Upgradation, 157–158, 189–190. *See also* Biometrics project management guide
Uptime, levels of, 276
User motivation, retinal recognition, 98
Username, 40
US federal government biometrics and KPI. *See also* Biometrics
  biometric data interchange formats, 23–26
  biometric technical interface standards, 28–29
  Common Biometric Exchange Format Framework (CBEFF), 26–27
US federal government biometric testing standards, 30–31. *See also* Biometrics

V

Variability
  interclass, 167
  intraclass, 167
Variable number tandem repeats (VNTR), 114
Vascular pattern recognition. *See* Vein pattern recognition (VPR)
Vector matching *versus* cohort matching, 146
Vein pattern recognition (VPR), 51, 69–75, 128, 129, 322. *See also* Biometric technologies
  advantages/disadvantages, 72–75
    acceptability, 73
    collectability, 72
    performance, 72–73
    permanence, 72
    resistance to circumvention, 73
    uniqueness, 72
    universality, 72
  components of, 71
  techniques used
    diffused illumination, 71
    direct illumination, 71–72
  Yarco Company (case study), 74–75
Vendor lock-in, 162
Verification, 10, 17
  and enrollment templates, 14–15. *See also* Biometrics
  matching, 146, 147
  templates, 14, 55, 258, 262
Vertical IT asset scaling, 272, 273
Virtualization attack, 299. *See also* Cloud computing
Virtual memory, 246
Virtual private networks (VPN), 232, 234–235
  and biocryptography, 261–262
  biometrics-based, vulnerabilities of, 255
  building, 241–242
  client, 233
  cost-benefit analysis, 235–236
  endpoints, 238
  firewall/server, 233
  implementing, 236–238, 249–251
  management responsibilities, 238
  mobile, 234
  security policy, components of, 238–239
  testing, 246–248
Virtual servers, 142, 286, 314, 315, 323. *See also* Biometrics in cloud
Virtual wrapper, 26
Voice recognition. *See also* Biometric technologies
  advantages/disadvantages, 102–103
    acceptability, 103
    collectability, 103
    performance, 103
    permanence, 103
    resistance to circumvention, 103
    uniqueness, 102
    universality, 103
  factors affecting, 101–102
  market applications, 103–104
  working of, 100–101
Voice recognition system, 38